The Giant Killer

Readers are encouraged to go to www.MissionPointPress. com to contact the author or to find information on how to buy this book in bulk at a discounted rate.

Published by Mission Point Press
2554 Chandler Rd.
Traverse City, MI 49696
(231) 421-9513
www.MissionPointPress.com

ISBN: 978-1-950659-47-0
Library of Congress Control Number: 2020904065

Printed in the United States of America

THE
GIANT
KILLER

DAVID A. YUZUK WITH NEIL L. YUZUK

MISSION POINT PRESS

Table of Contents

Introduction.. *vii*

1 Dark Skies.. 1

2 Worse News... 8

3 In the Beginning.. 14

4 Entering the Kingdom of Richard the Lion-Hearted..... 21

5 The Hunt Begins.. 25

6 A Silver Star is Born.. 29

7 The Giant Killer... 35

8 Lord of the Late-Night Lounges............................. 38

9 Reporting for Duty... 43

10 Heading to 'Nam... 59

11 Charlie Don't Like Visitors..................................... 68

12 The Silver Star.. 75

13 Snakes and Rats.. 92

14 Special Forces... 100

15 The RIF.. 105

16 A Girl Named Jane.. 109

17 The Seventies.. 117

18 Entrance to the Rabbit Hole.................................. 122

19 A Man Called Kates.. 138

20 Swamps and Gators.. 147

21 Razor's Edge.. 154

22 Undercover.. 158

23 A Man Called Griff.. 166

24 The Trial... 179

25 Back to the Abyss... 195

26 Beyond Love's Limit..................................... 200

27 Nightmares of the War................................. 209

28 On the Streets.. 212

29 The Last Week of Richard's Life.................. 230

30 Riddles in the Warehouse............................ 243

31 A Hero's Transformation............................. 254

32 Deeper into the Jungle................................ 256

    Documents and Photographs...................... 260

    Making the Documentary........................... 281

    Resources for Veterans.............................. 297

    Acknowledgments...................................... 301

    About the Authors..................................... 304

INTRODUCTION

*"Don't slide down the rabbit hole. The way
down is a breeze but climbing back's a battle."
—Kate Morton*

*"I have noticed that sometimes I frighten
people; what they really fear is themselves.
They think it is I who scare them, but it is the
dwarf within them, the ape-faced manlike
being who sticks up his head from the depths
of their souls.*

*"Most dwarfs are buffoons. They have to make
jokes and play trick to make their masters and
guests laugh. I have never demeaned myself to
anything like that. Nobody has even suggested
that I should.*

*My very appearance forbids such a use of
me. My cast of countenance is unsuited
to ridiculous pranks. And I never laugh.
I am no buffoon. I am a dwarf and nothing but
a dwarf." —Par Lagerkvist*

My name is David Yuzuk. As a veteran police officer I
can affirm there are no words more chilling to hear than,
"I have to tell you who I really am." This statement, uttered
by a tiny homeless man, changed my life forever. It would
be the last investigation in my twenty-year police career,
dragging me down the rabbit hole of CIA conspiracies and

stolen classified weapons. It would become a three-year journey of discovery stretching from the bloody jungles of Vietnam to the dangerous streets of Iraq and Venezuela, all in search of the peripatetic Captain Richard J. Flaherty. I would learn that not all of our teachers stand in front of a class or appear in the shape or image we envisioned they'd have. One of my teachers, seemingly forgotten by society, came to me later in life. The mantra he would incessantly preach was for me to discover my quest and live a life of no regret.

My name is Neil Yuzuk, and I'm David's father. I first learned of Richard Flaherty's existence when David called me and told me about this forgotten hero's life. My first job was to help research Richard Flaherty's story. As I became more involved in trying to uncover his mysterious life, I discovered Richard was a shadow — a wisp of fog who showed a different face to all he knew. Getting an answer to one question only created three more.

Although I never met him he became a part of my life, as did the brave men he served with, his family, friends, and the good people of Stamford, Connecticut.

I, too, became a student of Richard Flaherty.

———————————

This book was written as narrative non-fiction. The dialogue presented was sourced from countless recorded interviews of witnesses, family, friends, and acquaintances. The dialogue of Richard Flaherty was based on recorded video, notes, and my recollections of our conversations spanning over a fifteen-year time period.

The military events recounted in this book are based on more than twenty interviews with U.S. soldiers. The in-depth interviews of Vietnam veterans covered events recollected from a time period of over fifty years. Many of these events occurred under life-threatening circumstances; because of this perceptions and memory are often limited.

Secondary sources included a study of Richard Flaherty's personal journals, private letters, notes, VA medical records, travel documents, passport, previously published media accounts, and voluminous official U.S. military logs and histories. Those sources, credited where applicable, were utilized to provide a framework, to fill in gaps during periods when the primary sources weren't available, and to verify or elaborate upon the participants' recollections.

After the release of my documentary more people who knew Richard contacted me and provided me with information, filling in missing pieces of his life. I vetted all the folks who came forward in order to provide as accurate a picture as I could paint of Flaherty's life. Several names have been changed for privacy or security reasons, but all descriptions and information included about the individuals are accurate.

Note: All the following passages are written from David's viewpoint, except for the interview of Frank Sosa that was conducted and written by his father Neil.

THE GIANT KILLER

Dark Skies

"Sometimes it's the journey that teaches you a lot about your destination." — Drake

May 8, 2015
Aventura, Florida

As the afternoon's dark thunderstorm rolled in off the coast I backed my marked police car into an alleyway. The heavy raindrops hammered on my windshield as I searched through my papers for that phone number. It took me three days of digging — and a hell of a lot of favors — to get the number, and I didn't want to waste any more time, so I made the call.

A no-nonsense voice answered: "Hello."

"Hi. This is officer Yuzuk from the Aventura police department in Dade County, Florida. I'm trying to get ahold of retired ATF agent Fred Gleffe."

The voice warmed a bit. "Officer Yuzuk, I wasn't expecting your call so quickly. Look, I'm just finishing up some work in my garage. Is there something real fast I could help you with?"

"Yes sir. I wanted to check up on a man I've known for the last fifteen years who frequents my city. He mentioned that he worked undercover with you on a really big case."

"You got a name?"

"Flaherty."

He paused long enough for me to check my phone to make sure the call wasn't dropped.

"Richard J. Flaherty. How is old Captain Flaherty these days?"

"Homeless," I reluctantly replied.

Gleffe paused, cleared his throat and replied, "Homeless ... that's a damn shame."

"Yes sir, it is."

"Richard was one hell of an undercover operator, and we certainly couldn't have gotten that case started without him."

I was shocked by the revelation and jerked up in my seat. "So, it's true?"

"Absolutely. Look, give me a call tomorrow and I'll go over the whole case with you."

"That would be great."

"Anything else?" Gleffe seemed more relaxed and expansive.

"Yeah, one last thing. Richard kind of figured I would check up on him. He asked me not to contact you, or anyone else involved in the case."

"Why is that?" Gleffe asked.

"He said if I asked too many questions it could be bad for my career — and dangerous to his health."

"Well, it's been over thirty years. I'm not sure about how much danger he would be in, although that operation did piss a lot of people off."

"He said it had to do with classified weapons," I added.

After another long pause, Gleffe finally said, "Let me think about the case and we'll talk more tomorrow. How about I give you a call from my office landline ... let's say around five p.m.?"

I noted a slight change in his voice. It seemed a little less friendly, but I kept the thought to myself and answered, "Yes sir, that sounds good. I really appreciate the help."

In south Florida as quickly as the afternoon summer thunderstorms roll in, they just as quickly dissipate. On the bustling business street a large and imposing shadow emerges from an alleyway.

The man casting the shadow is sixty-nine-year-old Richard James Flaherty. Although Richard would always claim that he was every bit of four-feet-nine inches tall, his actual medical records reveal he was only four-feet-seven-inches.

Close to seventy years old, he still walks with a sure step; his posture is ramrod straight despite the fact he is lugging a heavy backpack. Under his faded navy-blue T-shirt and blue jeans is the taut ninety-eight-pound muscular body he's maintained since his teens. Although small his body is perfectly proportioned, leading most who see him to think he is an athlete of some sort, perhaps a jockey or gymnast.

Flaherty walks into the local Publix supermarket, which has served as his base of operations for the last ten years. He quickly selects a ham and turkey sandwich and pays for it at the front cashier. Several customers stare at the tiny homeless man, some with curiosity and others with disgust. Flaherty avoids all eye contact and mutters to himself as he exits the business. It's the same chant he's been reciting over a lifetime filled with the frustration of how people react to him: *What the hell do they know about me?*

As Flaherty prepares to cross the street he quickly scans the bushes for the tell-tale signs of an enemy ambush. He could always sense an ambush by a tingling sensation he felt rolling down his arms and into his hands. Choosing a safe path before moving forward was as natural to him as breathing. His last priority was to keep his head on a swivel and check his six to make sure those State Department agents weren't close behind.

Is it time to move on? Time to find a new area to melt

into? Am I still even on their radar? Of course I am ... I know too many of their damn secrets.

Although the State Department agents were always at the front of his mind, he wasn't all that sure about the helicopters he could glimpse flying off in the distance. He was positive they weren't standard police or news helicopters because the sound emitting from their main rotor blades was the distinctive, primal rhythmic thumping from a 1960s Bell UH-1 Iroquois Helicopter — better known as a Huey.

Flaherty heads toward the library, knowing that if the "government" ever needed his services again his point of contact would be a local public building where he could blend seamlessly into the crowd.

Flaherty walks inside the building, enjoying the feel of the refreshingly cool air conditioning as it surrounds his body. He inhales deeply, savoring the earthy almond smell the books give off commingled with a coffee aroma drifting from the staff's office.

He strolls over to a familiar aisle of books and hunts the crowded shelves for his target. Once located he stands on his toes and reaches as high up as he can to grasp *Paradise Lost*, by John Milton.

He takes the familiar book over to an empty corner desk and sits down with his back against the wall. He scans the room one more time until he feels it's safe, pulls out his sandwich and opens the book. Across the room, partially hidden in the shadows, a large man observes Flaherty.

As Flaherty is about to bite into his sandwich, his adversary bounds towards him and removes an "L"-shaped silver object from his front pants pocket.

Flaherty looks up from his book, but it's too late: his red-faced adversary is already standing over him, clutching an asthma inhaler. The inhaler emits a snake-like hiss into the obese library security guard's mouth as he depresses the trigger.

"This is the last time ... I'm going to warn you they don't want bums hanging out in here. Get your shit and get going."

Flaherty gathers his belongings without protest and exits the library. *Paradise Lost* is left open on the desk.

As the blazing sun starts to recede on the horizon another round of dark clouds rolls in from the ocean. Flaherty walks half a block towards his familiar shelter, a covered bus bench. He slides his heavy backpack off his shoulder and sits down on the metal bench next to a mother and her small daughter. Seeing Flaherty, the mother protectively puts her arm around her child and draws her close.

An elderly grizzled homeless man walks by, extending his gnarled hand to beg the mother for change. She turns her back on the man and continues to shield her daughter from a world of threats.

Flaherty rips his sandwich in half and motions to the older man. The man swiftly snatches up the sandwich before the gift can be retracted. With a glance over his shoulder the man hurries off to a familiar isolated alcove in the bushes. Feeling as safe as he possibly can, this hunger-worn societal outcast devours his sandwich with huge wolfing bites.

Flaherty sits patiently at the bus bench as people stream on and off buses and crowds walk hurriedly up and down the street, passing him without notice.

All the while time stands still for Flaherty. He sits motionless and alone as the day turns to night. The brake lights from the herd of cars sitting in traffic reflect off of the shelter's plastic walls, painting Flaherty in an eerie red glow.

The storm clouds once again hang over the city, and with a rumbling belch the sky opens up with heavy sheets of rain. The deluge seals Flaherty inside his shelter, further isolating him from the world.

May 9, 2015 — 1:15 A.M.
Aventura, Florida

Eight hours after my phone call to agent Gleffe the rain has stopped, and the streets are already bone dry. The parched earth has absorbed all the moisture it can, and Flaherty — after leaving his shelter — goes to sleep under his usual palm tree. He is sleeping in his customary upright position, with his backpack bolstered against the base of the tree for padding. Surrounding him are the dried crunchy leaves and twigs he nightly places around his perimeter to alert him if anyone comes near.

The faint sound of a helicopter startles him awake. Flaherty instinctually reaches for the memory of his Colt Python revolver holstered on his hip. As his eyes regain focus he scans the sky for the helicopter, only catching brief glimpses of a dark silhouette heading away from him. Flaherty stands up, grabs his backpack and heads off towards the empty city street. He stares into the cloudless dark sky, thankful for his most faithful midnight companion: the moon. *Finally, a break in the gloom.*

He cautiously crosses the street at the crosswalk and glances up several more times, looking for the helicopter.

A pair of car headlights slice through the dark empty streets. As Flaherty is about to step onto the median a bright white light appears out of nowhere, catching him entirely by surprise.

A loud explosion of shattering glass and twisting metal mixes with the heavy thud of a one-hundred-pound man being struck by a car traveling at twenty-five miles per hour.

Flaherty is lifted over the front left bumper — his head crashes into the car's A-frame and side window. The violent impact causes the metal A-frame to warp, the side window to shatter. Flaherty is then violently propelled upward and forward, his backpack separating from his body to flip through the air. He crashes and tumbles in a violent landing fifteen feet away on the dirt of the median. Flaherty's backpack rolls forward onto the edge of the street as papers escape from its ripped sides.

As the car drives away the street light reflects off a tiny bloody sneaker finally coming to rest upright in the middle of the street.

The sun is rising over the beautiful Turnberry Isle golf course as an athletic male and female jogger round a corner of the walking path that runs adjacent to the golf course. Several lawn mowers drone in the background, and the smell of fresh-cut grass permeates the crisp morning air.

The early rising BMW-driving Starbucks crowd is just exiting their shiny high-rise condos, calculating which route to the office will cause the least amount of suffering in Miami's three-lane parking lots — ironically called 'highways'.

As the jogging couple crosses the street the woman notices a child's sneaker oddly lying in the middle of the road. As she gets closer to the sneaker she looks into the median, seeing a bloody pant leg and foot jutting out of the bushes. The rest of the person's body is obscured by foliage.

"Oh my God!" wails the horrified woman.

2

The Worst News

"Sometimes the worst place to be is inside your own head." — unknown

May 9, 2015 — 6:39 a.m.
Aventura, Florida

The knocking was getting louder, angrier. I rolled over and barely cracked open an eyelid. 6:39 stared back at me in a taunting way. Maybe the knockings were those exploding nightmares you supposedly get when you're having a stroke in your sleep? I closed my eye again and twisted the pillow under my arm into a new position, hoping to get just a drop more rest. It had been another sleepless night, and the deprivation was causing me to experience bad headaches and a lack of patience.

Another round of angry knocking. *This better be some type of apocalyptic emergency*, I thought, *otherwise I'm going to be really pissed*. I slowly got up, disregarding my standard ritual of gradually stretching my back to avoid the heart-stopping back spasms (and accompanying trips to the ER for rounds of morphine and muscle relaxants).

I walked downstairs, being extra cautious not to aggravate the eggshell-like discs in my lower back. It's the inevitable price you pay for wearing a gun belt and ten-pound vest for your entire career. Another set of loud angry knocks reverberated off my tile floor and into my brain.

Is this ever going to stop? There're only two types of people who knock with that much urgency and tenacity: the guys from work when the proverbial shit has hit the fan, and my neighbors (who generally decide to bang on my door for faster service instead of dialing 911).

"Hold on, God damn it!" I yelled as I limped to the door. My good knee gave a satisfying pop, which alleviated some of its discomfort. It was probably just jealous of all the attention I was giving to my bad back and jacked up right shoulder.

I opened the door and saw a young police officer from the department. Clean shaven, uniform with creases that could cut your finger, and shoes shined like mirrors. *Yeah, he's new to the force*, I thought. *His face is just too smooth, it's a blank slate. I give him one more year before he's jaded and disillusioned, with dark raccoon circles under his eyes like the rest of us.*

My memory started to kick in even without the help of my three cups of caffeine. I'd met him once or twice before in the hallway, trying to keep up with his field training officer. I think I had coffee with him at a two-story parking garage we nicknamed the Bat Cave. He had this deadpan look on his face; I've seen this look before, and it never bodes well. He was sweating, but it wasn't from the Florida heat (though this lasts all year, except for maybe three days in mid-January). After twenty-some odd years of being a cop you can tell heat sweat from nervous sweat. This was definitely nervous sweat.

"This better be good … " I said, searching my brain for his name.

"Max. You remember? We had coffee once at the Bat Cave. We were also in that foot chase last month — you remember, after the bailout on the train tracks? Desoto had …"

"Right. Max. Good to see ya again. Can you make this quick?" I mumbled while I feebly tried to rub the pounding out of my temples. Last night when I did finally fall asleep something jolted me wide awake only an hour later. I'd spent the next few hours trying to corral that runaway horse in my brain, the one which stampedes at full gallop as one thought endlessly morphs into a million more.

The sun crawled up behind the rookie's shoulder to blast me full in the face, making my eyes squint and sharpening my already-fierce morning headache. They say sunlight travels ninety-three million miles through space to get to Earth. This morning it must have been hugely disappointed to travel so far only to land on my wincing, unappreciative face.

"Dave, you know that little homeless friend of yours, umm, I think his name is Bill?"

"Richard. His name is Richard. What about him? Someone try to mug him again, or did he get into trouble?

"No. He's dead."

Practical jokes can be good. Some of the guys I worked with were masters at getting the best of you and pulling strings. But some practical jokes cross the line. This joke crossed the line and kept running. I didn't lunge for him (which was my first impulse), though I could see by his body language he knew it was a possibility.

Let's try this again, I thought. "What did you say? Dead? Who put you up to this?" My voice got louder as my brain fog started to clear.

"No one, man. The Sarge told me to stop by your house and give you the info before I went home. I got stuck last night working mids and just about everyone is on the scene. I mean, I've never seen so many ... "

"Who's everyone?" I impatiently asked, my heart starting to race at the thought that this might be real.

"Dude. Everyone. The supervisors, all the guys, the Traffic Homicide squad, and about five different news vans. It looks like he was crossing the street and got hit by a car. A hit and run."

My heart pounded in my chest as the reality of Richard really being killed started to kick in. I got that sinking feeling in my gut, the one you feel when you lock your car door and realize the keys are still in the ignition. Or when you realize you really *did* leave the oven on when you're already driving down the highway — and have been for hours.

"Any skid marks?" I asked, trying to keep the bile from rising too high in my throat.

"I didn't see any. Oh wait, yeah, the Sarge was saying something about there being no skid marks. You see I was blocking traffic at the intersection, but I heard some of the guys talking. They said it might be ... "

"A drunk driver," I interrupted.

"Yeah ... how'd you know? Hey, you want to know my theory?" He started to walk into my house until he realized I wasn't moving my arm from blocking the door. "You got any coffee?" he asked as he continued to inch his way forward. He then placed his hand on my arm to push it away.

He stopped when he felt the rigidity of my arm and swallowed hard. I stared at him as I've stared at hundreds of suspects who were on my last and only nerve. Suspects whose final line of dialogue before I completely lost my shit was: "Officer, I don't know how come my son keeps getting all those bruises — he probably falls a lot." Or, "I only ran into the store for a minute. I cracked the window open for him — I didn't know it would be that hot today, I do love my dog!"

I eased myself back from the edge and thought about cutting the rookie some slack. I wasn't the sharpest knife in the drawer, either now or back when I was the FNG at the department, although I was certainly not as lost as this kid. Eighteen years just flew by in the blink of an eye — where did they go?

The sun was really cooking now, making me feel like an ant under a magnifying glass. I took a few deep breaths and licked my lips, planning my next move, thinking about who I'd call to confirm this.

"Where's the scene?" I asked.

"One nine-nine and Aventura Blvd.," he answered. Spoken like a true cop. Cops say each number individually — like one-nine-nine — instead of saying a hundred and ninety-ninth. Radio protocol and all, and it does sound a little cooler.

I'd got all the information I needed for the time being and gave him the obligatory, "Okay Max, thanks for the heads up. I appreciate it," along with a wave of thanks as I began to shut the door. I then heard Max's hand slap loudly on the door, stopping it from closing.

I gave Officer Rookie another 'What the hell are you thinking?' stare.

"Hey Dave, one last question. Is it true?" he asked nervously.

"Is what true?" I replied through clenched teeth.

"You know ... is it true he was a jockey? Everyone says he was."

"No. It's not." I started to close the door one more time. Dave going back to sleep: take two.

Another slap to the door. Another few hundred PSI on my molars as my jaw clenched. My dentist was going to have a field day on my next visit.

"Well, is it true that he's a millionaire? The guys are saying that too," he continued.

"No, Max. Thanks again for the info, and take care."

That was it. I shut the door and locked it twice. I could hear his muffled response coming through the door as I walked down my front hallway. I came into the kitchen and popped some ibuprofen in my mouth, knowing it would do absolutely nothing for my back — but maybe it could slow down this run-away headache.

I made my way back up what now seemed like fifty flights of stairs to my bedroom, only to find my five-pound fourteen-year-old chihuahua Pedro asleep on my side of the bed, his head on my pillow. *Will the indignities ever end?*

I gently pushed him over to his side of the bed; he protested with a grunt and reclaimed my side once again. I closed my eyes and tried to go back to sleep, knowing full well it was an exercise in futility.

I looked up at the ceiling fan and tried to register what I'd just heard. *Richard dead? How? He warned me not to look into that undercover operation. Shit, did I just get my friend killed?*

No skid or tire marks on the street?

The significance of tire or skid marks is crucial in any traffic crash investigation because speed, braking, direction, tire make, and vehicle models can all be determined from studying them. The significance of not finding any tire marks can also tell a story.

In almost all circumstances — whether the driver was sleeping, impaired, or even starting to slip into a diabetic coma — they will instinctively react after striking an object of significance. They do this by either hard braking, which would cause skid marks, or tugging the wheel abruptly in one direction or the other, which also causes visible tire marks to be left on the scene.

The only suitable explanation for no tire marks would be that the person was completely unconscious at the time — or that they weren't surprised by the violent collision and just continued on their way. The latter would indicate some level of intent: that the driver purposely and knowingly struck the object or person.

I picked up my phone and called the day shift sergeant. The call was brief and concise, but the news was confirmed. Richard was dead. They had no leads on the suspect. Everyone knew I was friends with Richard, and the sergeant gave me his condolences just before he barked some orders to a nearby officer. "Gotta go," he said. "Media." Then silence.

The phone dropped out of my hand and landed on my chest as I gave out a long painful sigh.

3

In the Beginning

"Three things cannot be long hidden: the sun, the moon, and the truth." — Buddha

My story with Richard Flaherty didn't begin that early morning of May 9th, 2015: it actually started closer to 1999. In September of 1999, as that fresh-faced rookie officer I can barely now remember or recognize, I transferred from the small Miami town of Surfside to the bigger and brand-new police department in the City of Aventura. As a young cop who wanted to change the world the move to a more prominent department with more opportunities only made sense.

In terms of fishbowls, it was equivalent to moving from a small carnival-sized fishbowl holding one sad fish to a ten-gallon Christmas tank special with its own gravel and enough space to keep the fish from cannibalizing each other — or so I thought at the time.

As a rookie starting at the lowest salary, working off-duty jobs in uniform wasn't an option: it was a necessity. Ramen noodles and Subway sandwiches were the daily diet until I started to pay off some of my police academy debt. The debt you ask? I thought police academy was free and trainees were paid? Well yes, that is true for some lucky people. I wasn't one of them.

When I started to apply to police agencies in the mid-90s not too many departments were interested in my services, so I decided to put myself through the part-time police academy night school. Yes, there is such a thing — or at least there was such a thing back then. It was a grueling ten month ordeal that cost me somewhere around six grand, which I wisely put on a high-interest credit card.

It was while working an off-duty job at the movie theater that I first saw Richard Flaherty. Richard was fifty-four years old at the time, and although he was extremely small he walked with confidence, as if he knew something the rest of us didn't. Richard kept his appearance well-groomed, and his usual dress was either jeans or khakis pants and a polo-type shirt.

Richard was a regular at the movie theater, which was located inside the busy Aventura Mall. After seeing him a number of times we started to acknowledge each other with slight head nods. Ironically the tiny green Anole lizards that rule South Florida also greet each other with head nods, so I guess we weren't all that much more advanced on the evolutionary scale.

The head nods eventually turned into casual, "Hey, how ya doings?" It probably took more than two years before I asked him his opinion on the movie he just saw. Our conversations for the next few years were cordial but very brief.

I started seeing Richard more and more over the next ten years, mostly walking on the streets because he stopped going to the theater. I later found out that the mall considered him to be a transient and banned him from the property. This, despite the fact Richard was always a paying customer.

Richard had his older brother Walter — who was a lawyer — write a letter to the mall threatening to sue, and the mall quickly rescinded the ban. Despite the mall's reversal, I don't think Richard ever stepped foot in that theater again. I never asked him about it, but I guess he was just too insulted to give them his business anymore.

―――――――

By the mid–2000s I started running into Richard more and more as he made his way around the city, usually lugging his large backpack or pulling a luggage cart. When I would stop to talk with him we always kept it pretty simple: news, politics, jokes. I never questioned him about his personal life, and I only once asked him if he needed some help getting off the streets. I think that's why he continued to talk to me. I never pushed him to discuss things he didn't voluntarily bring up on his own.

I enjoyed our little talks because he was sharp and witty, and always had an interesting take on things. He read newspapers every day and possessed an incredible wealth of knowledge of the world and its politics. There were also those rare days when his alcoholism got the best of him, causing him to mumble and slur his words.

I never judged him for that because I believed the only person he was hurting was himself. For the record, I never saw any traits or behaviors in the fifteen years I knew him that would lead me to believe he used any illegal drugs.

By the late 2000s we would start to meet for coffee or a sandwich during my patrol shift. The only strange thing was that every six months or so he would disappear from the area for a couple of weeks. I always assumed he went either north or south to the adjacent cities, just to get a different view of things and break up the monotony. By around 2010 he started sleeping mostly in one spot under a palm tree next to a bus stop surrounded by a Publix supermarket, the library, and a Kosher Kingdom grocery store.

It was around mid-April of 2015, during one of our

lunches at a Subway sandwich shop, that my real journey down the Richard J. Flaherty rabbit hole began.

As we sat facing each other, wolfing down our sandwiches, he suddenly said with sincerity, "Dave, I think it's time I tell you who I really am."

My first thought was, *Oh, shit. Here it comes. Is it too late to grab my sandwich and run?* Cops are admittedly a cynical bunch. You have no choice, as the job starts to mold you towards always thinking and expecting the worst. You go from believing *I'm going to have a great day at work* to *I hope my next prisoner doesn't have hepatitis and an exploding bowel problem when I transport him on the hour-long drive to the county jail.*

So, when Richard said what he said I immediately anticipated the worst. *Is he wanted? Is he about to confess to committing some unsolved crime? How will the next statement he makes affect my and his life?*

"I was the smallest man to ever serve in the military. I went to Vietnam with the Army's 101st Airborne, and then after attending Special Forces school I was stationed in Thailand."

His statement made no sense. It couldn't be true, but I politely asked, "You're saying that not only were you in the military, but you're a Vietnam vet?"

"Yup."

"And you were also in Special Forces? In the Army? Wouldn't that make you a Green Beret?"

"Yeah, that's right."

My mind wrestled with his outrageous statement, which of course couldn't be true. But he said it with such conviction: either he was telling the truth, or he was far more delusional than I ever realized. If a tiny four-foot-seven-inch homeless man (that you'd known for the last fifteen years!) told you out of the blue he was some great war hero, what would you think?

"Richard, I'm sorry to be skeptical — and I'm not claiming to be any expert on military regulations — but doesn't the Army have a height requirement?"

"Yes, a height and weight requirement. The regs state you have to be at least five feet tall and weigh at least one hundred pounds to be accepted. Well, it took me over three years of letter writing to my congressman Donald J. Irwin, but I finally got them to recommend a waiver for my height. My only other problem came up when I reported to the Army on a Friday for my physical: I only weighed ninety-three pounds. They gave me till Monday to make it to one-hundred, which I did. I've kept it on ever since."

Well, the little guy definitely has his story down, I thought cynically. *I wonder how long it took him to dream it up?* "Ever get wounded? Receive any commendations or medals?" I asked.

"I was wounded three times in combat, although I only received two Purple Hearts."

"Two Purple Hearts. Is that it?" I asked incredulously.

"Oh yeah, and I also received the Silver Star. And two Bronze Stars."

"I'm sorry, the Silver Star?"

"Yup."

My mind reeled with more skepticism. His statement was outrageous, but if any part of his story were true there would be some record of it on the internet. I was too uncomfortable to outright ask him for some type of proof to back up his claims. *If this is what he believes and it helps him get through the day, then why crush his delusions and make him feel worse?* I thought. He certainly never used any "stolen valor," claiming he was a vet to obtain benefits, that I ever heard or saw. Anyway, my shift would be over in a few hours. I could check on his story when I got home.

Richard then briefly told me he worked undercover in a federal operation in the early eighties. "But," he added, "better to leave out the details. Parts of that case are still classified."

This was a lot of information to digest on my lunch break. I thanked him for sharing his story — which I didn't believe — and went back to work. That night I almost didn't go on the internet because I had better things to do with my time than waste it on Richard's delusions. But, eventually I got around to it.

Some of the first things to pop up were news stories about the undercover case, which named Richard and referred to him as a Green Beret Captain. It stated that he was secretly working with the CIA in order to funnel weapons to the Contras in El Salvador and Honduras. I then saw an incredible old black-and-white photo from a newspaper article that showed Richard standing in his Green Beret uniform next to a very tall soldier.

I got chills looking at the picture because it verified that Richard was probably the most extraordinary person I had ever met. All these years wasted with small talk, when this man was a treasure trove of knowledge and world experiences! I almost couldn't sleep that night thinking of all the questions I now had for him.

I've been asked (many times, by many different people) why he chose this moment to tell me some of his secrets. I say 'some of his secrets' because Richard Flaherty was one of the most private and compartmentalized individuals I've ever met. I later learned he kept almost all his professional and personal life from even his closest family and friends.

The little intricate pieces of his life that he did share were only different aspects (or chapters), so no one could truly see the whole picture. Everything was on a need-to-know basis, and I guess Richard felt that nobody really needed to know. I have my own theory on why he chose this time to reveal his past, and will delve into those reasons later on.

As I drove to work the next day I kept thinking about what an incredible underdog story Richard's life was — going from the heights of being a highly trained Special

Forces Captain and war hero to sleeping under a palm tree. This was a story I felt I needed to share with the world.

How could we as a society let veterans like Richard fall through the cracks and become homeless? Although I'd never made a documentary I felt this would be the perfect subject. Why not give it a try? As long as Richard agreed to participate, of course.

I found Richard later that afternoon under his tree reading a book. I pulled my squad car over and hopped out to talk to him. When he saw the big smile on my face he said, "I guess you found out I wasn't pulling your leg."

We made some small talk before I flat-out asked him if I could make a documentary about his life. He wasn't thrown off in the least bit and asked me, "If we start this project, no matter what happens you promise me that you'll finish it?"

"Richard, you have my word that to the best of my abilities I will do everything in my power to complete the project." Although we didn't have a Bible to swear on I gave him my word. There are many things in this world I'm not good at, but giving my word and keeping it (or being a "stand-up guy") is something I've always prided myself on, no matter what the cost.

Richard then added another stipulation. "Oh, and one last thing. If they ever make a movie about my life I want Brad Pitt to play me, and my quote is three million."

"Deal," I said. We shook hands, which was something new — once when I tried to shake his hand he'd drawn it away defensively. He later told me he didn't shake hands anymore because of old war injuries and his arthritis. I took care not to squeeze his tiny hand, but he did give me a firm handshake. That's how this whole project got started.

4

Entering the Kingdom
of Richard, the Lion-Hearted

Kosher Kingdom Bench

That night as my shift was ending I wanted to meet up with Richard and talk a little more about his life story. Richard wasn't at his usual palm tree, so the next place I knew to look for him was in his Kingdom: the Kosher Kingdom, a grocery store across the street next to the Walgreens pharmacy with a well-lit bench.

The bench sat in the corner of the small strip shopping lot. The "L"-shaped businesses provided ample cover for Richard's back, thereby denying enemies the opportunity for a sneak rear attack. The area lighting coupled with the height of the bench allowed Richard a tactical higher ground advantage: he could survey from a distance any threat that approached him.

I pulled my police car into the empty parking lot directly in front of Richard, exiting to talk to my friend.

Richard, as always, sat majestically on his bench, as comfortable and familiar as a king on a well-used throne. Instead of a royal adviser by his side he had his large faithful black backpack comfortably seated next to him.

His uniform of the day was his favorite blue jeans with a red T-shirt and a gray hooded sweatshirt. He cherished that hooded sweatshirt, which he sometimes draped over his head like a monk's cowl, allowing him to easily blend into shadows as he traveled the sometimes-dangerous streets. More practically, it also kept the rain and cold weather off his head.

Wisps of short gray hair barely covered the top of his scalp, while several small white gauze bandages sat in an arc on the crown of his head. A full gray goatee fiercely hung from his face, adding to his medieval appearance. His stage lighting was provided by a distant street lamp commingled with the red glow of the Kosher Kingdom neon sign.

As I approached Richard he stood up on the bench. At this height, we were now eye level. He pointed his finger into the air as if he was about to perform a monologue on a Broadway stage.

Richard dramatically proclaimed, "Every man's life eventually comes to an end. It's only the details of how he lived and how he died that distinguish one man from another."

"Mark Twain?" I guessed.

"My Irish ass Twain. It's Hemingway, of course."

"I finally finished *For Whom the Bell Tolls*. Nice read."

"Nice read? My boy, that's a classic to be savored like a bottle of Remy Martin Cognac on a cold winter's night. I have another assignment for you to think about if you're up to the challenge."

"Another 'what's the sound of one hand clapping' riddle?"

"No, my boy. This one is straightforward. What principles do you live your life by? Hmmm? Everyone has a place in their heart they would never betray — what about you? What would you never betray?"

I blinked. "You want that answer now?"

"Just food for thought. Take your time on that one, because it's one of the keys to the journey of self-exploration."

Richard's joking manner seemed to quickly evaporate as he sat back down on the bench. It was replaced by a sullen demeanor.

"You know Dave, for a big guy in all the years I've known you I never felt that you were looking down on me. I appreciate that."

Before I could get out a thanks he smiled a mischievous grin and boisterously asked, "Now my boy, what do you want to know?"

I pulled out my pen and notebook. "Can we start at the beginning, and you kind of walk me through your life?"

"Well, since it seems my date Jennifer Lopez can't make it tonight, why not?"

As Richard started to recount his tale his reflection in the store window served as a mirror and filter, revealing his view of the world. The constant buzzing of the Kosher Kingdom neon sign served as the soundtrack to Richard's incredible story.

When I later checked my watch, I realized almost two hours had magically passed. The only thing that would look strange to an onlooker watching us talk was my relentless pacing. This rhythmic dance was something I had to do to stand for long periods without my back locking up. Even with all my precautions I still added an extra pain killer to my nightly regimen to offset the hours of standing on concrete.

For the next two weeks Richard and I would meet up periodically. I would listen to him, taking notes as he remembered new details of his life. My idea at the time was to gather as much information as I could, then see if I could interest any real filmmakers in helping me document Richard's story. I certainly had no experience making a documentary — and how could I possibly pay for it? The plan was to learn as much as I could, then wing it from there.

I thought — just like we all do with people in our lives — that I would have a lot more time with Richard. I felt no sense of urgency to collect as much information from him as I could, as fast as I could. It never dawned on me that our time was running out; in less than ten days he would be killed.

The only video I ever shot of Richard Flaherty was taken that night at his kingdom. It was a grainy two-minute cell phone video of Richard starting to tell his story. Oddly my cell phone battery quickly lost power and I had to end the video abruptly. I should have realized then it was an omen of things to come.

5

The Hunt Begins

After Richard was killed I had no interest in making a documentary. Even if I did, how could I continue with him gone? All I had were some disjointed notes and a couple internet articles. I felt like a sailboat stranded dead in the water. Richard, my wind, was gone and I felt hopeless — my idea of getting his story out to the world was over.

The open homicide case was assigned to veteran traffic officer Harvey Arango. Traffic unit officers are specially trained in all traffic-related investigative fields. I'd known Harvey my whole career, and I've always respected him as an officer. He was a low-key, easygoing giant of a man, a devoted husband and proud father.

One of the first investigatory pieces to the enormous jigsaw puzzle of Richard's life was his storage unit. I only remembered the storage unit because Richard mentioned it in passing one day when I was taking notes. He said he might have some military documents in there that he could show me. Since I was the only person in my department who knew about it, I suggested to Harvey that it would at least be a place to start looking for leads into his homicide. Harvey agreed; he had to wait on red-light camera video to be analyzed, along with the forensic analysis of small fragments of the suspect vehicle left on the scene. Road patrol officers would continue to canvass the area by knocking on doors and asking if anyone saw or heard anything suspicious the night before.

Harvey and I then responded to the storage unit warehouse, which was only a mile outside of Aventura. When we arrived we made small talk with the desk manager Julia Mirabel, who sadly told us Richard was a friend who used to talk to her when he stopped in — she always enjoyed their conversations. Harvey then asked Julia for copies of Richard's billing statements. She printed them out and handed them to Harvey. Harvey sat down at a small front lobby desk to review the statements.

As Julia leaned on the lobby counter a dark shadow suddenly came across her face. "Oh my, I forgot about his cat."

"Richard had a cat?"

"Well, not really his cat. There's a one-eyed stray cat that lives behind the building next to the train tracks. A couple of years ago some kids had the cat trapped under a car and they were throwing rocks at him. Not everyone likes strays, you know. Richard scared them off, and ever since then the cat always greeted him. They say once a cat becomes feral it can never go back to living in a home. I guess it's a trust thing. Well, every time Richard came by he always fed that cat a can or two of tuna fish — not the cheap kind, but like Starkist. Richard was the only person who could ever get near it."

"Would you do me a favor and feed ... " I started to say.

"Of course I will. You don't even have to ask. It's the least I could do for Richard."

Harvey finished scanning the pages and motioned to me, indicating he was ready to look at Richard's unit.

The inside of the warehouse reminded me of that last scene in *Raiders of the Lost Ark*, where the wooden crate containing the Ark is brought into a huge warehouse holding god knows how many secrets. Dim fluorescent lighting and musty smells welcomed Harvey and me to the world of human storage. We walked down aisles of hundreds of individual units, packed to the ceiling with the contents of people's lives; it conveyed a somber feeling, like being in the Roman catacombs.

We eventually located Richard's unit, also fully packed to the ceiling with items. There was a large metal filing cabinet, dozens of medium-sized cardboard boxes, briefcases, luggage carts, suitcases, backpacks, fishing tackle boxes, and metal tool kit cases.

Inside these were thousands of legal, military, and research documents, dozens of handwritten letters and notes, pictures, books highlighted with Richard's notations, reams of small legal pads filled with cryptic writing, and several mysterious items whose secrets I would later uncover. It seemed Richard left me all the bread crumbs I needed to research his life.

I began to feel a slight breeze gently pushing against my sails. A force was starting to prod me towards fulfilling my promise, the vow I made to Richard to finish what I started. I wasn't just an investigator trying to solve a murder: I felt more like a searcher, seeking not only the answers of how Richard died but (more importantly) how he lived.

Harvey and I began the tedious task of removing all the items from Richard's unit, hoping to find some clues. One thing which immediately became apparent was that, although Richard did a great job saving these thousands of documents, they certainly weren't organized in any order whatsoever. Every box down to each scrap of paper had to be sorted out.

We dragged over several large tables, and for the next few hours quietly went about trying to organize the chaos. As the piled-up boxes on the table grew higher the morning quickly turned into late afternoon. Just then Harvey got a call on his cell phone that would change the entire investigation. I watched his face transform from business as usual to plain shock.

After hanging up the phone Harvey turned to me and said, "You ain't going to believe this shit. Sarge just got a call from the Miami-Dade Police Department Homicide Unit. Their sergeant says he has the hit and run driver at his station — and the suspect wants to talk."

"That was quick," I answered.

"That's not the crazy part. The crazy part is the suspect is actually a Homicide Unit employee. I'm heading over there now to do the interview."

Harvey grabbed his things and left me alone with the huge pile of items — and even more questions.

6

A Silver Star is Born

Where do you begin the story of a man's life? *In the beginning,* I thought as I considered what my first move should be. I would fly up to Richard's home town of Stamford, Connecticut to speak with his two cousins Donna Marlin and Jeanie Rinaldi, who were the only family members he kept in contact with. I also set up meetings with several of his childhood friends.

Guarding the mouth of Galway Bay in western Ireland, with its three-hundred-foot cliffs and unusually savage weather, are the isolated Aran Islands. On this rugged, windswept landscape studded with ancient Celtic forts Richard's grandfather, Joseph Patrick Flaherty, was born in 1883. By the age of sixteen his restless spirit drove him to leave his beloved islands and travel to Ireland's mainland.

Speaking mostly Gaelic, he stopped in the small village of Feakle, which was nestled alongside the Sliabh Aughty Mountains — considered the loftiest and wildest land in all of Ireland. Local legend states that the village was founded when the Patron Saint Mochonna lost his tooth in the soil and decided it was God's message for him to stay and build his church. It was there Joseph Flaherty met his future wife, Bridget Smythe. In the early 1900s Joseph and Bridget Flaherty left their home country to start a new and prosperous life in America.

Rounding out Richard's colorful pedigree, his grandfather Vincenzo James Rinaldi was born in 1872 in Calitri,

Italy. "Jimmy" was an infamous character in the 1920s New York City scene, known for the wild saloons he operated and his alleged ties to mobster Al Capone.

Nestled on the southwestern tip of Connecticut and just a stone's throw from the shining waters of Long Island Sound sits the city of Stamford. Once known as Rippowam by the Native Americans, this coastal town's population swelled in the 19th century when New York residents started to build their summer homes on its picturesque sandy shoreline. Blue collar families followed *en masse* when industrial factories started replacing the agricultural facilities. The new train routes also made commuting into Manhattan more economical. Known as the birthplace for the electric dry-shaver industry, the Schick Company employed almost 1,000 workers by the 1940s.

On November 28, 1945, Richard James Flaherty was born in Stamford Hospital. Richard's mother, Beatrice Rose, didn't know at the time of his birth that her blood type was Rh-negative. The Rh factor is a protein that can be found on the surface of red blood cells.

If the blood of a Rh-positive fetus gets into the bloodstream of a Rh-negative woman, her body understands it is not her blood and fights it by making anti-Rh antibodies. These antibodies can cross the placenta and destroy the fetus's blood. This reaction may lead to serious health problems — and sometimes even death — in a fetus or newborn. The complication only occurs with second-born children, which Richard was; therefore his future was sealed before he took his first breath.

It was further believed that the Rh complications caused a hormonal imbalance in his system which stunted his growth. Medically speaking, Richard would be considered a dwarf. Dwarfism is described as a medical or genetic condition that results in an adult height of 4' 10" or shorter. More specifically Richard would be referred to as a Proportionate Dwarf, meaning his arms, legs, trunk, and head were proportionally the same as an average-sized person.

Richard's father, Walter Timothy Flaherty, Sr., worked for Sears Roebuck as a salesman, and his mother Beatrice Rose was a housewife. They owned a two-bedroom house on Twenty-One Park Street, which at the time was considered a suburban middle-class neighborhood. Richard's older brother Walter Jr. (or "Timmy" as most people called him) was known as a serious and studious young man, who eventually became a lawyer.

Richard's childhood friend Rick Farina (who still lived in Stamford) invited me to his home to talk about his memories of Richie Flaherty. In a strong New York accent he told me:

"Growing up in Stamford in the fifties and sixties was like a page out of small-town America. Nobody locked their doors, and our parents didn't worry about us getting kidnapped or attacked when we played outside. I grew up in the house across the street from the Flaherty's, and me and Richie were best of friends.

"Seemed like almost every day me, Richie, and a couple of other neighborhood guys would head to the park and play football or baseball all day long. Then we'd head to Tony's Drive-In to buy fifty cent hot dogs. Life was simple. Richie, besides his size, was just like any other kid in the neighborhood growing up. I mean we all knew he was self-conscious about his size, and how couldn't he have been? When we went into a store people would stare at him. It would make me uncomfortable, so I can't imagine how it made Richie feel. They might have stared, but nobody that I can remember ever picked on Richie — at least not when he was around me.

"The cool thing about when you'd go to the Flaherty's house was they always had all sorts of animals. Dogs, cats, rabbits, you name it. I remember Richie's dad at one time was breeding dogs. I believe they were Boxers, and he would sell them to make some extra money. I guess that's where Richie's love for animals came from. He used to tell me back then that he wanted to be a veterinarian.

"There was a really sad incident that I remember, and it still bothers me to this day. Anytime I smell anything remotely close to the smell of street oil or tar it immediately brings me back to that time. You see, back then in Stamford twice a year they would repair our roads by spraying this strong-smelling black oily substance, and then they'd cover it with sand. I guess it was meant to seal the asphalt, but when you'd drive over it for a day or two till it dried it would stink to high heaven and stick to your car tires. They'd usually spray it in the middle of the night.

"It was sometime in the fall of 1956. My mom just had surgery and me and my dad went to the hospital to pick her up. My dad had bought a powder blue 1955 Chevrolet Bel Air, and he was upset because the sand and tar were ruining the undercarriage. As we got closer to my house that smell was just overpowering. My street was always the last to get sprayed down, so it would also be the last to dry. As we were pulling up to my driveway something small darted into the road. It probably misjudged its timing because its feet were hampered by that fucking tar, and before my dad could swerve the car we hit it. My dad rushed my mom into the house because the smell was making her even sicker. He told me to go see what we hit. I was only ten years old and I didn't want to go, but I did anyway.

"It was rolled into a ball and covered in that vile tar, and I knew it was dead. I recognized it immediately: it was JJ. A couple of weeks earlier Richie's cat Virgil gave birth to a litter of three kittens. The smallest in the litter — the runt — was Richie's favorite, and he named it JJ. I picked up JJ's tiny lifeless body and walked it over to Richie's house. I went into his back yard and up onto his porch, because back then we never used each other's front doors — you'd just go to the back door and yell out your buddy's name. That day I didn't call out to Richie, I just quietly knocked on his door.

"Richie answered the door and I watched his face drop. It was the first time I ever saw my friend emotional. Even as a kid Richie was reserved — he was what you would call stoic. I mean he would laugh and joke with the best of them, but I never saw him lose his cool or cry or anything like that. If it's possible to see a person's heart break I saw it that day when I handed him JJ. He was devastated, but he blamed himself for allowing JJ to get of out the house. It was the only time I ever saw my friend cry. We put JJ in a shoe box and buried him together in his backyard, and after that we never talked about it again. I always wondered if the smell of that foul black tar affected Richie as much as it always did me?

"By the end of junior high school me and Richie started drifting apart. I was playing baseball in high school, while Richie focused mostly on training in karate and boxing. I remember many times coming home from baseball practice and seeing him in his garage breaking wooden boards or punching bricks to toughen his hands. He would do it for hours alone, hardening himself up. It's kind of weird when you look back at it now, because it seemed Richie was preparing himself for some type of mission or duty that he would later be called upon to fulfill.

"Richie's love of alcohol also started at a young age when he discovered his parents' hidden liquor cabinet — obviously that habit would follow him for the rest of his life. He was a practical joker and would stop at nothing for a laugh, but he also had the ability to laugh at himself. Our buddy Dennis Connors told us that when Richie slept over at his house he would throw his pillow and blanket in the bottom drawer of Dennis' dresser and then go to sleep inside it. I mean Richie was a real character, and he was always up to something."

After speaking to Rick Farina I headed over to Richard's old high school Trinity High. One of Richard's high school teachers now retired met me in a hallway holding a copy of Richard's old high school year book. He opened the book to the page with Richard's picture and remarked, "Richard Flaherty was a rambunctious young man who made a lot of the nuns who worked at the school rethink their vows." Next to the picture of Richard was the caption, "The giant killer... clever and witty... newspaper... college bound."

"Can you tell me how he got the nickname the giant killer?" I asked.

7

The Giant Killer

1959
Trinity High School Boys Locker Room
Stamford, Connecticut

Inside the raucous locker room of Trinity High, a group of students quickly change from their PE outfits back into their school clothes. Fourteen-year-old Richard Flaherty, standing in front of his locker, takes his time and waits for the room to clear. Once the last student leaves he stands on his toes and grabs a large brown paper lunch bag off the top shelf. He opens the bag and removes a small bottle of vodka, a syringe, and an orange, placing them on the bench next to him. He deftly uses the syringe to remove ten ccs of vodka from the bottle and then injects it into the heart of the orange. He replaces the items in the bag as he hears another student enter the locker room.

The student is Flaherty's nemesis Rex Simms, AKA "T-Rex" (due to his large body and short arms). Simms turns the corner like a shark prowling the water looking for an easy meal and menacingly struts over to Flaherty.

"Well well well, if it isn't my favorite dipshit dwarf. What'ya doing there, Richie boy?"

Flaherty, without answering, hastily shuts his locker and heads off in the opposite direction, lunch bag protectively tucked under his arm. After a few steps a large hand grabs his shirt and pulls him backward.

"Hey dwarf, I'm talking to you." Simms slams Flaherty hard up against the lockers and stares down into the little student's eyes with a wicked smile. Instead of fear Simms infuriatingly sees Flaherty coolly and unblinkingly staring back up at him. A coach walking by looks in their direction and shouts, "Cut that out, you two. Get to class." Flaherty uses the opportunity to scamper out of the locker room and run down the hallway. Right before he reaches his classroom he ducks into the janitor's closet.

Once inside he opens his lunch bag and pulls out a dozen smoke bombs. He nimbly laces the fuses together and places the smoke bombs in the farthest corner. He turns back to the door and peers outside, making sure the coast is clear. He heads back into the corner, pulls out a Zippo lighter and lights the long fuse.

Flaherty rushes out of the janitor's closet and enters his classroom. He scrambles to his seat, seeing that T-Rex made it to the classroom before him. Simms, with the fury of unfinished business, glares back at Flaherty.

A cloud of black smoke starts billowing out from under the janitor's closet door into the hallway. Inside the classroom Flaherty opens his lunch bag, removes the orange and starts peeling it. The teacher is about to start her lesson when the fire alarm starts ringing. Flaherty calmly starts to eat his vodka-laced orange with a knowing smile.

The teacher says, "Okay everyone, you remember our drills. Line up. Rex and Richie, like we practiced: grab the flag and lock up both doors." Flaherty and Simms both nod in unison. The teacher lines the students up and leads them out of the classroom. Flaherty goes to the corner of the room and picks up an American flag. Simms stands by the door, waiting with a key. The last student exits the room as Simms places his key in the door.

Simms barks at Flaherty, "Hey moron, let's get ... "

Flaherty — wielding the heavy metal pipe used to hold up the flag — swings hard into Simms' knee. Simms screams and collapses to the floor, writhing in pain.

The teacher ended his story by looking at me with a devilish grin and added, "after word got out about Simms, Flaherty's new nickname was The Giant Killer and nobody ever picked on him again."

8

Lord of the Late-Night Lounges

Kosher Kingdom Bench

"After high school I didn't have any direction, so I took a few classes at the community college," Richard said.

"The classes didn't help guide me, so I took the first job I could get: sweeping floors at a mechanics shop. Not all that exciting, but my weekends were different. They were all about boozing, chasing women, and causing trouble. Usually I did all three at the same time." Richard chuckled, then spoke in a quick montage of fuzzy memories that I could easily picture in my head.

Richard — drunk and rowdy in a crowded pub — jumps up onto the bar and starts kicking drinks like an NFL field goal kicker, as a raucous crowd cheers him on . . .

Richard is sitting at a booth in a lounge with several girls, animatedly telling his big stories. The table is filled with empty glasses, and Flaherty jovially waves the waitress over for more rounds . . .

Richard rolling around on the floor of another local pub, locked in a fist fight with a drunken patron. The crowd slides over to give them space, enjoying the pugilistic entertainment . . .

Richard ending most of his weekends by either being thrown, pushed, or carried out of several different watering holes around Stamford.

Wooly Bully by Sam the Sham and the Pharaohs blares from the jukebox as Flaherty relaxes on the concrete, leaning against a trash bag in the alleyway behind the Ole' Irish Pub. Having been thrown out of his second bar that night, he sits in the rear alleyway trying to recover his bearings. His shirt is torn, his lip is bleeding. *But it was a good scrap*, he thinks to himself, as he gave as good as he got.

The beefy forty-six-year-old bartender Chuck Wilton locks up the side door of the business and walks over to Flaherty. Chuck looks down on Flaherty and asks, "Think next time you could find an even bigger guy to fight?"

Flaherty looks away in anger. *This guy is always eyeballing me. Always cutting me off when I'm not even close to being drunk. He better mind his own business.*

"C'mon Rich, I'll give you a lift," Chuck says as he helps Richard stand up. Richard starts to push the man off him, but Chuck with a deadpan face shakes his head and adds, "Not going to happen, don't even try."

Chuck guides the woozy little man towards his car — parked nearby — and puts him in the rear seat. Richard uses the opportunity to lay down and close his eyes as Chuck drives off. While driving Chuck turns down his radio and takes a glance back at Flaherty. He frowns and asks, "You ever stop and think why you never get in trouble — or even barred — from all the places you've been thrown out of? Hmmm? Why everyone buys you drinks no matter how obnoxious or belligerent you get?"

Chuck looks over his shoulder again at the reclining Flaherty, waiting for an answer. Flaherty looks away from Chuck's stern gaze with disdain. Chuck adds, "Well, I'll tell you why. Because people don't expect anything more out of you. They think this is the best you'll ever be."

Chuck now has Flaherty's full attention. He angrily stares at the back of Chuck's head. After a short drive, Chuck pulls the car over and cuts off the engine. He exits the car and opens the rear door for Flaherty. Flaherty sits up and looks around to see they've arrived at Stamford's Veterans Park.

"I don't want to be here," Flaherty grumbles.

Chuck grabs Flaherty roughly by his shirt and pulls him out of the car into the park. They stop at the base of a newly erected plaque. The plaque reads: Master Sergeant Homer L. Wise, Medal of Honor, June 14, 1944, United States Army.

Chuck stares at Flaherty and somberly says, "You understand this is hallowed ground that we're standing on? Someday Homer's statue will stand here."

"So what," Flaherty disdainfully answers.

Chuck shakes his head in dismay. "I couldn't believe it when I heard Homer was your cousin. He's one hell of a man. Being related to him must mean living in a mighty big shadow."

Flaherty looks away, uninterested.

Chuck continues, "You know that Homer never went into a battle without first reciting a passage from the Bible?"

"How could you possibly know that?" Flaherty disbelievingly asks.

Chuck lifts up his shirt sleeve to show Flaherty a tattoo on his forearm. It reads Company L, 36th ID.

"Because I was with him in 1944, on that hillside in Italy. Me and half the guys from our platoon wouldn't be alive it if wasn't for him. It's a small fucking world, huh?"

"He's never talked to me about the war," Flaherty grumbles.

"Don't suppose he would. There are things about war a

man never talks about. Look Richard, if you want people to stop pitying you you have to first stop pitying yourself."

Flaherty looks towards his cousin's plaque for the first time.

Chuck continues, "Don't you want more out of your life? Don't you ever dream about doing something incredible?"

Chuck puts his hand on Flaherty's shoulder. "Son, there's a lot of people in this town that think you'll never amount to anything, but I think different. Homer single-handily took on a German tank and saved a bunch of our lives that day." Chuck looks him squarely in the eyes as he adds, "I challenge you to do better!"

"And do what, fight? Go to Vietnam?" Flaherty fires back. "The Army don't allow dwarfs, and even if they did it's not my fight." Flaherty shrugs off his hand and storms away, swaying drunkenly.

"Richard, there's something else I need to tell you!" Chuck shouts after him. Flaherty refuses to stop, walking all the way home in the middle of the night.

Kosher Kingdom

"The next day people in the neighborhood were buzzing with some sort of news, but I didn't really pay attention to it. It wasn't till Monday morning, when I saw the Stamford Advocate newspaper, that I learned my childhood friend Richard Broadhurst was killed in Vietnam. I knew him from grade school, and he was a real easy-going quiet kid back then. In my teens I used to run into him at the beach where he was a lifeguard and we'd just say hello.

"He was the first man I ever heard of from Stamford to be killed in 'Nam, and he wouldn't be the last. All told by the war's end we lost twenty-nine men. A week later they had the service for him at Gallagher's Funeral Home, and there was a huge turnout. It literally looked like the entire city was there.

"You hear people talk about life-changing moments. Well, that was mine. When I saw Richard's mom and dad and all his brothers and their sorrow, the war finally became real. Before that Vietnam was barely ever talked about. Now everyone knows all about Vietnam — they've seen it depicted in hundreds of movies and books. They all know names like Saigon and Ho Chi Min, but back in 1965 it just seemed like a make-believe place. A place your high school guidance counselor would threaten to send you if you didn't get good grades, almost like some type of boogeyman. Like a lot of other guys my age I didn't buy it. It didn't affect me or my world, so why should I care?

"To tell the truth I didn't feel sadness that day. All I felt was shame. Here I was running around like the world owed me something, and there was my friend lying in a casket. For whatever reason he joined the Marines, he went, he served, and he died for something that mattered. I now had a better understanding of what Chuck was trying to get through my thick skull.

"After the service as I watched them shovel dirt onto his coffin, I swore I would no longer be satisfied with just getting by. I would become something. I would prove them all wrong, and I would figure out a way to get into the military. I would then outwork, out-think, and out-compete any man that got in my way."

9

Reporting for Duty

"We sleep safely at night because rough men stand ready to visit violence on those who would harm us." — Richard Grenier

Army Basic Training is hard no matter what year you go through it, and back in 1966, with a war raging on the other side of the world, the urgency for creating the perfect fighting man was at its peak. The Army would either mold that new recruit or break them in the process. The individual meant nothing because the team meant everything.

Flaherty started his physical training several months before entering the Army. He received his exercise regimen from his childhood friend James Connelly, who was already working as an athletic trainer. The two started their days with five-mile runs in Stamford's Cummings Park, ending with a quarter-mile sprint on the beach in the sand. The next two hours were devoted to weightlifting and calisthenics. In those few months Flaherty built his body from a thin ninety-two pounds to a rock solid one-hundred-and-three pounds of pure muscle.

October 10, 1966
Fort Jackson, South Carolina
Basic Training

Fear of the unknown is on full display as a greyhound bus full of fresh-faced recruits pulls into the Army base at Fort Jackson. The earlier bravado and all-around grab-assing during the two-hour ride comes to a halt as the bus passes under the foreboding Welcome to Fort Jackson sign. Located near the state capitol of South Carolina, Fort Jackson is the largest and most active Initial Entry Training Center for the U.S. Army. Created in 1917 and named after Major General Andrew Jackson (AKA "Old Hickory"), Fort Jackson served as the newest Army facility to train fighting men during World War I.

It is a crisp October morning. As soon as the doors slide open a drill instructor explodes into the bus, barking orders faster than the startled young men can comprehend. With every unintelligible word firing out of his mouth the angrier he becomes. No one really understands the drill instructor at first, but some words are made very clear thanks to the punctuating finger he jabs at the end of his shaking hand. "YOU FILTHY MAGGOTS GET OFF MY BUS.... NOW!" The spit from his mouth flies in all directions to spatter the young men, who are too scared to even wipe it from their faces.

Stiff with anxiety and confusion the men awkwardly try to make their way off the bus, moving like a herd of frightened antelope. The only exception to the panic is the driver, who sits quietly in his seat, enjoying this rerun of the show he's seen a dozen times before; waiting to see which one will start crying, and who will freak out.

Flaherty is also watching the show and taking notes as he shrewdly sizes up his competition. He leaps off the bus and runs to the vertical yellow inspection line, an uncanny feeling of calmness and belonging flooding his senses. The bellowing colorful abuse the drill instructor spouts at the hapless young men is nothing compared to the lifetime of insults and ridicule he's already endured. These instructors are experts in mental warfare, and Flaherty relishes the challenge.

Because the new recruits are so attentive to the drill instructors's barking instructions they are given permission to comfortably relax in the front leaning position, also known as a push-up. As most of the men struggle at the count of thirty Flaherty is just getting warmed up.

Flaherty and the other recruits are marched into a long green building. There, a noncommissioned officer assigned to uniform detail stands slack-jawed, staring down at Flaherty.

"You gotta be kidding me!" screams the NCO from behind the uniform counter. "Where the heck am I going to find a uniform to fit you?"

Later that afternoon on the football-field-sized tarmac, Sergeant Baker (assigned to train the recruits of Delta Platoon) stalks over to the young men already lined up in formation. *Here we go again — let's see what this year's crop yields!* After taking a deep breath he walks down the formation line, carefully scrutinizing the fresh-faced recruits with their newly buzzed haircuts. After eyeing several six-foot-tall soldiers, he stops dead in his tracks: the next man in line barely reaches to the last soldier's shoulder.

Staring down at Flaherty Baker mutters, "What the heck is this?"

"Private Flaherty, Sergeant."

"Son, what on God's green Earth are you doing wearing that uniform?" Baker looks around the base suspiciously. Is he being pranked?

"Sergeant ... ."

"This is not the goddamned Marines, you don't say Sergeant here boy. You will address me as Drill Sergeant." Baker scratches at his head, still not buying into it. *This someone's idea of a joke?* Loudly he barks, "At ease, men! Boy, you follow me." Baker turns sharply on his heel and marches toward the officer's headquarters, Flaherty following close behind.

Stopping outside of the captain's office, Baker turns to Flaherty. "If this is legit — who the hell did you con? The Army don't allow anyone under five feet tall, and you ain't no more than four foot ten."

"Four foot nine to be exact, and I promise you, Drill Sergeant, this is no con. It took me dozens of letters and phone calls to get my congressman and a three-star general to finally recommend me."

"Is that so?" Baker sneers. Turning, he knocks heavily on the captain's door.

Captain Addams' Office
0700 Hours

Flaherty is standing at attention in Captain Addams' office. Addams is seated at his desk, smoking a foul-smelling cigar and looking through paperwork while Baker stands rigidly across from him.

Baker is unable to contain himself. "There's no way sir, this has got to be a screw-up. For Pete's sake, this kid has no chance of even lasting a day out here."

Addams, still reading, holds up a hand. "Sergeant, hold on ... ."

Baker interrupts, "I'm sorry sir, but if this isn't a joke then you got to transfer him somewhere else, because he's going to get someone hurt."

Flaherty, without permission, pipes in as he turns to Baker, "Give me a chance, Sarge. Throw everything you got at me, 'cause I can take it. You ain't going to break me and I'm not going to quit."

Addams puts down his paperwork and laughs. "Little son of a gun does have a set of balls. His paperwork seems to be all in order. Look, Sergeant, I agree with you," he turns to Flaherty, "and son, no disrespect, but you really don't belong here." He turns back to Baker and states, "But if you can't break him you're stuck."

Baker, without waiting to be dismissed, pivots and stalks out of the room. "We'll see about that," he barks over his shoulder.

———————

At first his fellow recruits didn't say anything to Flaherty, or comment about his size. The fear and shock of the first few days were too much for them to focus on anything else. Everything moved at a breakneck pace. A typical day consisted of an early rise at 0400 hours, a five-or-six mile run, and then breakfast. After breakfast there would be training at either the rifle range, the sand pit for hand-to-hand combat and calisthenics, or the obstacle course. Then lunch, then off to more running. That was how it went, non-stop day-in-day-out under the watchful, unblinking eyes of the drill instructors.

As days turned to weeks, everyone in the company became more at ease with the routine and more comfortable with each other. Nicknames and joke-filled insults began to fly in the barracks. No one was immune from the razzing, and Flaherty took more than his share of dwarf and leprechaun jokes.

Charlie Skaggs was the only recruit who openly objected to Flaherty being in the platoon. Skaggs was a scrapping six-foot tall two-hundred pounder from the steel mills of Pennsylvania. Working in the mills was all the Skaggs family knew until the mid-fifties when the industry, plagued by soaring costs and declining profits, started massive layoffs. Skaggs had a chip on his shoulder starting on day one, and now he had Flaherty in his sights as his target.

––––––––––

One night in the barracks before lights out Skaggs makes his case in front of the other recruits. Pointing an accusatory finger in Flaherty's face he chides, "How the fuck is private dwarf going to carry any of us if we go down in the 'Nam? Huh? I say he needs to try the Boy Scouts before he thinks of becoming a soldier. And if he don't want to go I say we give him some help."

Flaherty listens to Skaggs' speech without reacting, instead getting himself ready for bed.

A couple of other recruits also have their doubts about Flaherty's effectiveness in combat, but they keep their opinions to themselves. They, like most of the other men, are more concerned with getting through basic training than worrying whether some tiny guy belongs or not.

The following morning the drill instructors enter the barracks for an inspection. Flaherty's bed and footlocker are always in perfect condition; he confidently stands at attention in front of his bunk. Sergeant Baker is about to move on to the next man when he decides to check the interior of Flaherty's footlocker. All of Flaherty's uniforms are crumpled into piles, and in the corner is food from the commissary, which is strictly prohibited in the barracks.

"What the fuck is this, private Mighty Mouse?"

Flaherty calmly looks over his shoulder and acknowledges the infractions.

"Well, I'm waiting for a fuckin' answer."

Flaherty says nothing and stands stoically at attention with his eyes forward. Baker, looking toward Drill Sergeant Amos, yells, "Sergeant Amos, you take private Flaherty and administer a little disciplinary physical training by running him into the dirt until he explains why he would disrespect the Army and my hospitality by allowing him the privilege to live and train at Fort Jackson."

"Private Mighty Mouse, get your ass out the barracks and run as fast as you can to the range, so we can become better acquainted," barks Sergeant Amos. As Flaherty sprints out of the room, Skaggs (standing in front of his bunk at the back of the barracks) allows himself a slight grin.

Five hours later Sergeant Amos follows Flaherty back into the camp. Flaherty's uniform is soaked in sweat and dirt, but his head is still held high and his spirit isn't broken.

"Goddammit Flaherty, your uniform is a disgrace. Go get yourself cleaned up and squared away, then get some chow."

"Yes, Drill Sergeant."

Amos eyeballs Flaherty as the little man makes his way back into his barracks. Amos then walks over to Sergeant Baker, who stands watching.

"Well?" asks Baker.

"I ran that little fucker up and down every hill on this base, and he wouldn't tell me who messed up his locker. Stubborn little prick," Amos says with a crooked smile.

"Look Bob, I know you and some of the other guys are starting to take a liking to him, but honestly you got to ask yourself: is he going to be able to hack it in 'Nam?"

"Tommy, why don't you cut the kid some slack? We've both seen those G.I. Joe squared-away-looking mother-fuckers who crack once the rounds start coming in."

"Got to disagree with you on that one, Bob. And for the record, the captain is also on board with me helping Flaherty decide this isn't the right place for him."

Sand Pit
1300 Hours

The following day the recruits line up in front of Sergeant Baker at the sand pit for their hand-to-hand combat training. Sergeant Amos, along with drill instructor Gabe Walker, stand by on hand to supervise the drill.

"Gentleman, today will be the first day you start training with a partner on the techniques we've taught you. I don't want to see any of that sloppy bar brawling, and remember gentleman — this is just training, so don't kill each other. Now partner up."

Flaherty turns to the man next to him, private Jimmy Bernard. Bernard (who is one of the smaller men in the platoon) sighs with great relief. Baker watches the men pair up and decides to make some changes.

"Bernard, why don't you get with private Gonzalez and Mr. Flaherty — partner up with Mr. Skaggs." Flaherty without hesitation runs over to the smiling Skaggs.

"Now gentleman, we will start out easy with a little warm-up drill where I only want to see standing grappling with transitions into hip tosses. Once your partner is down, you will start on your feet again. You will go on my whistle, and you will only stop when you hear the whistle blow twice. Is that clear?"

"Yes, Drill Sergeant!" the platoon roars back in unison.

Baker walks over to the other instructors with his whistle in hand as the nervous recruits size each other up. With a cruel smile Skaggs says to Flaherty, "I'm going to snap that little neck of yours." Flaherty doesn't respond, but focuses on the task ahead. Baker sees Amos disappointedly shaking his head.

"Sorry, Bob. It's time to separate the men from the boys."

With that Baker blows his whistle. Flaherty, without hesitation, launches himself straight at Skaggs in a blinding series of kicks and punches. The three sergeants gape as they watch the little wolverine-like man attack his grizzly bear opponent with a ferociousness they've rarely seen before.

Flaherty purposely targets Skaggs' knees, groin, eyes, and any other vulnerable areas on his body. Amos finally comes to his senses and yells, "Hold it hold it. Everyone stop!" All the recruits stop their exercises except Flaherty and Skaggs, who are locked in a no-rules fight.

Baker grabs Amos by the arm, shaking his head. "They'll need to work it out themselves, 'cause I ain't blowing that whistle."

The platoon now surrounds Flaherty and Skaggs as everyone — including the instructors — are caught up in watching the surreal moment.

Flaherty throws a quick three-punch combination as a distraction to get Skaggs to focus on his hands. Flaherty uses the opportunity to hook his heel around Skaggs' foot, neatly tripping him to the sand. Skaggs can't believe he's now on the ground, seeing Flaherty looking down at him; with the metallic taste of blood in his mouth he staggers back to his feet and grits his teeth. The fight isn't over just yet.

After the initial surprise of Flaherty's attack, Skaggs is finally able to grab the whirling little buzz saw that's been chewing him up. He now easily rag dolls Flaherty, lifting him high in the air and slamming him down again and again into the ground. The recruits scream encouragement to the underdog Flaherty.

"Fuck him up, Flaherty!"

Flaherty is eventually able to break out of Skaggs' grip, but the damage is done, and Flaherty is limping badly. Skaggs, whose nose is now leaking like a faucet, uses the opportunity to launch a series of big looping punches. Flaherty covers up as best he can and tries to roll with the blows, but the one-hundred-pound deficit is apparent. Skaggs' punches make loud thudding noises as they bounce off Flaherty's shoulders and ribs. One huge punch gets past Flaherty's guard and knocks him flat on his back. Skaggs, breathing deeply with blood dripping down his face, looks down on his opponent. "You had enough?"

Flaherty spits out a wad of sticky blood from his mouth and says, "I ain't even started yet." With that Flaherty — still on the ground — kicks upward to the inside of Skaggs' knee. Skaggs is knocked off-balance as his knee hyper-extends backward. The two gladiators continue their back-and-forth brawl for the next two minutes.

Baker watches disbelievingly, never thinking that Flaherty would last more than thirty seconds. As the battered and bloody combatants exchange punches and kicks Baker blows his whistle twice and steps in between the two men.

"What do you think, Mr. Skaggs? Has he earned your respect?"

Skaggs rubs his jaw and tests a loose tooth in his bleeding mouth. Out of breath, he answers,

"Hell yeah."

"Okay, you two take a breather on the bench." Baker watches the two men limp away, then turns and calls out to the platoon.

"Bring it on in and take a knee." The platoon surrounds him in a half-circle and falls to a knee.

"You sorry-looking motherfuckers better have learned something from those two. You guys think I'm wrong for allowing them to fight? Hmmm?" Baker stares into each man's eyes, awaiting a reply that he knows will never come.

"No? Nobody? I know what you're all thinking. Y'all thinking that I'm a cruel son of a bitch, and that that wasn't a fair fight. Now get this in your heads, war ain't a fair fight! It's dirty, it's nasty, and you have to do whatever you gotta do to win. Killing another man in war isn't about honor or nobility — it's a matter of killing him before he kills you. It's that simple."

As Flaherty and Skaggs walk towards the bench,

Skaggs asks, "You think you could teach me some of those kicks?"

"Sure, if you promise not to bounce me off the ground!" The two men shake hands and sit down.

After Flaherty's fight, the men of Delta platoon looked upon him as their fiery little mascot. He was now their talisman, exemplifying the platoon's fighting spirit. The platoon would sometimes carry Flaherty on two of the men's shoulders as they ran in formation, Flaherty proudly perched in the air holding the platoon's flag, waving it in disdain at the other platoons. Many times a drill instructor from another platoon would use Flaherty as an example to admonish some of their slackers. "If Delta Company's Mighty Mouse can make that five-mile run wearing a seventy-pound rucksack you maggots damn well better smoke that course in record time!"

Army Obstacle Course
0900 Hours

Addams is standing in a muddy field next to Baker, watching a squad of new soldiers trying to get over the wooden wall. Most of the soldiers are just barely able to jump high enough to grab the top of the wall. In the background, a drill sergeant screams a profanity-laced tirade of encouragement to help motivate the men.

Addams, full of skepticism, asks, "And you're telling me he's able to make it?"

Baker shakes his head. "Just wait and you'll see."

"But how can he ... " Addams starts to ask as Flaherty's squad rounds a bend in the road and Flaherty is leading the pack.

As Flaherty approaches the wall, without breaking stride he pulls a screwdriver from his pants pocket, sharpened down to look like a reinforced ice pick. With the tool in hand he deftly springs as high as he can onto the wall.

His momentum drives the spike end deep enough into the wall to bear his weight. He pulls himself up with one arm and quickly grabs the top of the wall while removing his modified screwdriver. Addams shakes his head in disbelief. *Looks like a god damn tree frog skating up a tree.* Once sitting on top of the wall Flaherty replaces the screwdriver in his pocket, hops down, and keeps running the course.

"Well I'll be damned. Do the regulations allow that?"

"No one has ever tried it before, so there's technically nothing in the regs to disqualify him. Little bastard always finishes in the top five."

"And all the other PT requirements?" Addams asks.

"I first tried running him into the ground, and even though his goddamn rucksack weighs more than him — he wouldn't quit. He outperforms his entire squad in calisthenics, he can literally do push-ups for hours. Hand-to-hand combat he gets tossed around quite a bit, but always gets back on his feet. I thought he would've gotten murdered in the pugil sticks, but he's got the agility of a spider monkey — no one can hit him with a clean shot. Little half-pint made a believer out of me. I say he stays."

Addams shakes his head and adds, "Unfuckingbelievable."

March 1, 1967
AIT School
Fort McClelland, Alabama

Flaherty— in his third month of Advanced Individual Training — is lying in his cot looking through his mail, and opens a letter from the Army. He quickly scans it and a big smile comes across his face. The letter reads, "Congratulations — you are officially accepted to apply for Officer Candidate School."

Getting the offer for Officer Candidate School (or OCS)

is another massive accomplishment, because Army regulations state all officer applicants have to be at least five feet four inches tall to apply.

Flaherty once again relied on his brother Walter to start up a letter writing campaign to have that rule waived. This time Walter went even higher up the political chain in reaching out for help. Finally, U.S. Senator Thomas J. Dodd (father of Senator Christopher Dodd) stepped forward and assisted Flaherty in receiving the waiver. The waiver did not mean acceptance, it only allowed him to apply for the school through an interview process.

———

The interview is conducted in an empty wooden barracks, causing sounds to echo weirdly around the room. As Flaherty enters the room he's met by a board of three stone-faced officers seated behind a heavy worn metal desk that looks like it might've seen some combat in its day. Flaherty is motioned to sit on the one solitary wooden chair in the room, ten feet away from the desk. A fluorescent light hangs above him, buzzing intermittently and giving off a yellowish hue.

The officers ask a number of questions, then pass around his file to review his record. They huddle in deep discussion while Flaherty sits upright and stares straight ahead. The ranking officer, Captain Jake Scott, asks one last question of Flaherty before he is dismissed. "Private Flaherty, let's just cut to the chase. There's a reason why we have strict guidelines on height requirements for all our leaders. For effective leadership, the troops have to respect the man they are entrusting their lives to as he leads them into battle. Our other concern is, we don't want any officers in the position of power to abuse it if they had, let's say, a chip on their shoulder or a Napoleonic complex. My question to you is, how would you win the men over and become an effective leader?

And how would you handle the adversity and jokes about your size?"

Flaherty takes a moment to think, then launches confidently into, "First, I would never ask any man to do anything I wasn't willing to do first. I believe leaders, especially in a time of crisis, need to step to the front and lead by example. But in times of stability, ease off of the reins and allow the men to rely on their training and instinct in order to build their own self-confidence. I also consider leadership to be a privilege to better others, as opposed to using it as an opportunity to just satisfy my ego.

"And about my size ... I've faced a lot of kidding about my stature. I don't take it to heart, I just let it roll off my back. If they persist I'd ask them to step outside and handle it like a man while telling them that I'm eighty-five pounds of muscle, fourteen pounds of dynamite, and one pound of shit-kicking Uranium-238!" Flaherty's voice booms as he delivers his last line, causing it to echo around the room.

Flaherty's answer causes a halt to all paper shuffling and note writing. Shocked looks from the panel aim at Flaherty, but he doesn't flinch. Alone in his chair, he relishes the challenge and stares back into each officer's face. Captain Scott is the first of the three to break his composure and let out a hearty belly laugh. The other two officers, unable to restrain themselves, chuckle and shake their heads.

———————

On February 8, 1967, Flaherty started his six-month Officer Candidate School tour at Fort Benning, Georgia. Ken Deats — who attended that same class — remembered 'that tiny little guy Flaherty.'

"I remember when I first saw him at "Benning's School for Boys" inside the barracks and we'd all walk around in our skivvies — you know, our boxer shorts. Now as shocking as it was to see him in a uniform — that by the way never fit him — but to see him in just his skivvies you really could see just how small he was. I mean when we'd do these rifle drills he would have trouble swinging his rifle around because it was almost as tall as he was. But I never heard him complain, and he did everything we all did and then some.

"There was this one time we were marching over the parade ground and another company was marching the opposite way. The men in the other platoon started ragging on Flaherty, pointing and laughing, and one guy screams, 'Hey why the hell did they let a midget in the Army?' Flaherty didn't hesitate and screamed back, 'Because I got brass balls.' He was defiant, he was cocky, and that's what I remember about Flaherty!"

September 3, 1967
Jump School
Fort Benning, Georgia

Inside the crowded C-119 transport plane (better known as the "The Flying Box Car") two rows of jumpers (known as "sticks") wait nervously for their turn. Flaherty and the other troops are given the order to stand up and hook up. They secure the snap hooks from their main parachutes to a long wire strung down the length of the plane.

Jumpmaster Sergeant Stone diligently walks up and down the plane, conducting safety checks on his men's equipment as the old aircraft shudders and bucks. Stone mutters to himself the old jumpmaster adage — that it's scarier to fly in a C-119 than it is to jump out of it.

Stone walks over to Flaherty and rechecks the duct tape job he performed earlier. Flaherty is the only soldier with machine gun parts taped to his legs. Stone shouts over the roar of the engines, "Look, son, I'm not sure if that's going to be enough weight to keep you from drifting off, so remember to use all the techniques we taught you."

Stone turns to the first man in line, who is standing at the open exit door, and shouts, "Go!" The man hesitates, frozen in the door until another jumpmaster provides him the proper assistance by booting him out of the plane with his size twelves. Stone screams to Flaherty (who's next in line) the command: "Stand in the door!"

Like every other jumper on board, Flaherty is carrying two separate parachutes. On his back is the main parachute, known as the T-10, and in front of him is a smaller reserve chute in case things go bad. Below his reserve chute is his rucksack, which for most men hangs at knee level but on Flaherty hangs to his feet. Flaherty struggles forward awkwardly, baby-stepping towards the door as the rucksack bounces up and back off his lower shins.

Stone turns to the other jumpmaster and yells over the wind, "We jerry-rigged the shit out of his equipment, but who knows what the hell's going to happen? He might be blown all the way to Hawaii."

Stone sees Flaherty has his hands on each side of the door; he's properly leaning forward. He screams the order, "Go!" Flaherty fearlessly leaps out of the plane at twelve hundred and fifty feet above the Earth and is blasted backward by the wind.

One thousand ... Two thousand ... Three thousand ... Four thousand ... SNAP! Flaherty's canopy pops open, his fall jerks to a stop, and everything becomes quiet as he enjoys his seemingly slow descent.

10

Heading to 'Nam

Kosher Kingdom Bench

Richard paused for the first time in his story to stand and stretch out his own back. I'm not sure if it was my constant pacing and groaning or memories of his own back injury from a training incident that also had him performing the chicken dance. After a quick twist and a satisfying loud pop from his back, he sat back down and resumed his story.

"Let's see ... after graduating OCS and Jump School they sent me to Fort Campbell, Kentucky — home of the 101st Airborne — to continue training 'til I got my orders. In December of '67 I was shipped to Vietnam, which was just in time for the Tet Offensive.

"We started out in Bien Hoa, a major air base in the south that had almost everything you'd find back in the States: an officer's club, mess hall, PX. You could almost forget you were out-of-country until the nightly mortar and rocket attacks reminded you this was no vacation. While there I was given my first platoon leader assignment and put in charge of the 3rd platoon of Charlie Company, 1st 501, 2nd Brigade.

"To be a platoon leader in charge of an airborne unit was really something special because paratroopers were all volunteers: that meant I had the top-notch soldiers. As a twenty-three-year-old officer I was considered an old person in Vietnam. There were guys as young as seventeen in my platoon. Shit, we later found out there was a fifteen-year-old kid who doctored up his birth certificate to get into the Marines.

"So you've got all these young guys, and they're armed to the teeth, they can kill everything in sight if they want to. And you have to control that, so they fight and kill when you want them to; otherwise all hell broke loose. Because if somebody starts shooting in a situation, what'll happen is they'll all start shooting, and then next thing you know they're shooting everybody in sight. You don't want that. You need to be able to direct that fury where you want it to go.

"That was just one of the challenges of being a platoon leader, but it was the most important because you had the power of life and death. If you said somebody lived, they lived. If you said they died, they died. It was just like being in the Roman Colosseum.

"When we first got to Vietnam, they gave us this long briefing and said, "Well, you might see the enemy once every couple of weeks or so; you may not see them for long periods of time." Well, a couple of weeks later the NVA and VC launched the Tet Offensive, and everything changed real fast.

"When Tet started, the balloon went up everywhere. The Tet Offensive was intended to get the South Vietnamese to basically turn to the North and say, "Okay, we're going to rise up with you and rebel, throw these Americans out and take over."

"Every military unit in-country was engaged in a fight. That's when they flew us north to the base at Phu Bai, which was near the Imperial City of Hue. Hue was once the capital of Vietnam, and it was an ancient city surrounded by a huge wall. Built like a medieval castle with its citadel surrounded by ramparts and moats.

"By the time we arrived Hue was already under assault by two North Vietnamese and one Viet Cong regiment. There were probably around eleven thousand enemy troops massed in Hue at that time. It was still the early days of Tet and the Marines were in there fighting it out.

"We started operating around the outskirts of Hue: the only way I could describe it was like stepping into the bowels of hell. There were civilian and NVA bodies lying and rotting everywhere you looked. The fighting was so intense nobody was even thinking of recovering bodies. The sounds of gunfire and explosions were non-stop, and the volume was absolutely deafening. You could feel the explosive concussions constantly vibrating through your body and shaking your internal organs.

"Working on the outskirts of the city, we were engaged in one of the most dangerous of military maneuvers — trying to contain a trapped and desperate enemy. Actually, a pretty large contingent — maybe a whole enemy regiment — eventually escaped over the Citadel Bridge, crossing the Perfume river into the mainland. We called it the Perfume river because in the fall tropical flowers would drop into the water and float downstream, wafting fragrant smells onto the banks of the city. But the only thing I remember smelling during that time was the pungent rotting stink of dead people.

"After the NVA escaped they blew up the bridge behind them. They figured the Americans wouldn't be able to continue chasing them. But shit, we didn't let that stop us. We crossed the river and were right on their heels. We spent the next thirty days chasing large North Vietnamese units that were retreating out of Hue and other towns in the area. Most historians will say the battle of Hue was the bloodiest battle in all of the war, and on that I would have to agree.

"Our platoon and every other company would put out ambushes every night, and the NVA couldn't help themselves from getting caught in our traps. It was like throwing a net into a pond overflowing with fish; although these weren't regular fish in our nets. They were a lot nastier — more like piranha.

"Right around then was the first time I was wounded in combat. I got hit with some shrapnel from a grenade. We were stopped in a Vietnamese village when some sappers busted through our perimeter. Sappers were basically Vietnamese kamikazes — soldiers with explosives on their backs that ran head-first into your perimeter.

"They killed two of our guys on one position and then started to infiltrate our whole unit, but we eventually got them out of there and killed close to ten or fifteen of them. The next day we decided we weren't going to wait to be hit, we'd bring the fight to them.

"So, in the morning as soon as dawn broke we started in pursuit of this NVA unit. We did this for most of the day into the early afternoon hours, humping through some of the worst terrain imaginable, until we eventually came into this small deserted village.

"Even though it was deserted it still had the strong rotting oily fish smell of their favorite sauce, Nuoc Mam, mixed with the peanut buttery smell of their chewed betel nut. The villagers chewed the betel nut, which supposedly had a caffeine effect, and it also caused their teeth to turn a reddish black. There was always a corner inside their huts dedicated just for spitting the juice into.

"My platoon was the lead element. When we came into this village I started noticing AK-47 rounds scattered around on the ground — just like somebody was throwing out bread crumbs for us. We tracked it for about a hundred meters until I got this bad feeling and halted the platoon. I knew something wasn't right because we were following a trail they wanted us to follow.

"I radioed my company commander and told him that I thought it was a trap. He agreed, and ordered us to move off the trail and over to the end of the village to set up a defensive firing position.

"I then ordered everyone in my platoon to fix bayonets to their rifles. I said, "Before we get into our firing positions probe everything in the village. Probe every haystack, every sack of rice, and check for any kind of booby traps. Also look for spider holes, anywhere they can hide — because we don't want them behind us."

"If you think fighting an enemy waiting in front of you in ambush is bad, that's nothing compared to Charlie sneaking up your ass. Charlie was what we called anyone who looked even remotely suspicious in 'Nam. So, after probing everything we moved to the edge of the village near this big hedgerow; that's where I had my guys set up in firing positions.

"My company commander raised me on the radio and advised that he'd ordered First Platoon to leapfrog us, and for us to make sure our fire support was ready for them. So, we're trying to set up our fire support, but the area is too flat — we had no way of getting elevation over First Platoon, which would soon be in front of us. Tactically it was really bad, but out there you had to improvise on the fly and make do.

"Despite the disadvantage First Platoon goes through as ordered and the first thing they do is recon by fire. By indiscriminately firing their rifles and launching M-79 grenades ahead of them they were hoping to lure the enemy into giving away their position, tripping their ambush before it was sprung.

"Well, those M-79 explosive rounds must have got one of those NVA soldiers nervous. He started shooting before he was supposed to shoot; First Platoon was only about forty yards into the field when that soldier let loose with his AK-47. It's a good thing that happened, because if First Platoon moved another fifty yards forward they would've walked straight into the killing zone. So, the early warning gave them a chance to hit the ground and dive behind cover. Even so, a lot of those guys got hit because the NVA still had the jump on them.

"Even back where we were we started taking casualties. The amount of enemy gunfire coming in was so heavy it was like a wall of lead slamming into us. In the first ten seconds of the battle I lost my machine gunner; like losing an essential piece on a chess board before the game even starts. We needed to get some fire on that heavily infested enemy tree line, because they were chewing us up. Without my gunner it was nearly impossible.

"I radioed the lieutenant running First Platoon and I asked, "What's your situation?" He screamed, "I'm hit I'm hit and I got at least half a dozen people down out in this field."

"So, I grabbed a few guys and we ran up a trail that was along the right side. There was a little depression at the end of the trail where we would at least have some partial cover. We hauled ass down that trail, jumping into the depression as the bullets chased us.

"I jumped back and forth several times from the depression up to the trail, firing on fully automatic. Every time I would jump back into cover this NVA heavy machine gun would open up on us. Once I jumped up to start firing and an NVA round actually hit my rifle. I had the feed tray up and my rifle in front of me; the rifle went flying back and slammed into my helmet, making a loud thudding noise. It knocked me flat on my ass.

"I then hear one of my guys yell out, "The Lieutenant's dead!" As I'm lying dazed on my back I actually saw the bullets slowly coming through the bamboo toward me. It was like one of those slow-motion movies: I saw the bamboo actually splitting, and I'm going 'holy shit, I'm seeing even the smallest of details on that bamboo as it splits.' The next thing I did was to start to feel around my body, thinking, "Where am I hit?" And I wasn't. I remember actually saying out loud to myself, "No, you're not hit, you're fine."

"I looked over to my radio operator lying next to me; they'd shot the shit out of his radio. It had three big bullet holes in it, but somehow he was unhurt. Just stunned. So

I said, "I'm going back to get the other radio." Well, when I jumped up that machine gunner opened up on me again.

"He had me dead to rights and centered me with a burst, but I was just super lucky to dive at that exact moment. One round went through the cargo pocket of my pants leg. As soon as I hit the ground a grenade blew up next to me, knocking me sideways several feet. I didn't feel any pain whatsoever until I tried to run and felt the searing agony of several pieces of large shrapnel lodged in my leg.

"The adrenaline kept me moving, and I brought back the other radio. We continued to fight and we stayed out there, because when you're in a fight — it's not like a John Wayne movie, you know? It doesn't go all silent so you can talk to the sergeant, tell him how you're feeling and about the girl you left behind. There were no breaks, no time-outs 'cause I'm hurt. You had to fight, so we kept fighting.

"I was all bandaged up by this point. I don't even remember who bandaged me up, but most of my guys were pretty bad off. We then started to drag all our equipment and wounded out of there so we could call in an airstrike. After pulling back I laid on my back in the wet grass, totally exhausted. Looking up I could see those jets come screaming in, straight for the NVA.

"The NVA knew our tactics and tried to make a run for it out of the tree line. They were trying to run across a rice paddy to escape, but it's very hard to run in a muddy rice paddy, especially when you have two F-4 Phantom jets chasing you with mini-guns. That's the last I saw of that battle. I was placed on a stretcher and loaded into a Huey helicopter for evac. There's something special about those helicopters — something rhythmic and alive. I don't think you can understand how much that Huey meant to us in the jungle. It was life and death itself.

"I was in the hospital for about three weeks before I was sent back to my platoon. The doctors wanted to sideline me for six weeks, but every day I kept telling them that I was fine. I wanted to go back to my unit because that's when rumors started spreading that we were going to parachute jump into North Vietnam. I wanted to be back in time to be a part of it.

"I think I went to the 22nd Surgical Hospital in Phu Bai, then to the 67th Evac in Qui Nhon. Then I went to 6th Convalescent Center, which was on Cam Rahn Bay. You get your initial surgery — the meatball surgery — at the surgical hospital, then after a few days they move you to an evacuation hospital and you get another surgery done there. Then you go to the convalescent center to heal up and go swim in the ocean, because the more salt water you get on the wound the quicker you heal.

"I had this huge metal zipper on the side of my leg. Nowadays they use staples, but back then they used stainless steel wire. They would run it through and then twist it, so you just had a bunch of wire twists all up and down your leg.

"I had like eight or ten of these wire twists in my leg. All I could think about was getting those things out of me because they'd catch on damn near everything you walked by and hurt like hell."

Kosher Kingdom Bench

As Richard recounted his memories his cadence would pick up speed, the words spilling out of his mouth so fast I missed a lot of the detail. I needed him to slow down, give me less of a history lecture and more of what he was actually seeing and feeling at the time.

I told him we should take a quick break and grab something next door in the twenty-four-hour Walgreens pharmacy. I got us both two big bottles of Gatorade and a bag of beef jerky to quiet the rumblings in my stomach. Richard didn't want any food, which wasn't out of the ordinary

— besides the lunches we occasionally had I never really saw him eat.

By the time we got back to his bench he seemed a lot more at ease, as if getting the first part of Vietnam off his chest had calmed him down.

As he started up again his descriptions became more vivid, and he even began recalling the names of the men. Instead of sounding like a sterile lecture on Vietnam it now became more personal, visceral, and real. It felt as if I was walking next to him in the jungle ....

11

Charlie Don't Like Visitors

April 5, 1968 — at 0800
Eight Klick Ville

Airborne's Lieutenant Rick Lencioni checks the gear of 2nd Platoon one last time before leaving the landing zone — better known as LZ Tombstone. He is supposed to receive ten new replacements to bolster his platoon back to its normal operating strength of thirty-four men, but he knows this isn't going to happen. He's run his platoon on missions with sometimes as few as eighteen men, but that's Vietnam — you make do with what you have.

He is grumbling to himself about the wisdom of his boss's decision to send his platoon — along with Lt. Flaherty's 3rd Platoon — back into the Eight Klick Ville for another search-and-destroy mission. The eight-thousand-meter-long village bisected by a river is a known Viet Cong stronghold, and Charlie doesn't like visitors in his neighborhood. Lencioni's platoon already cleared the Ville two previous times in the last month, taking several casualties, only to abandon it back to the enemy. *Here we go again!*

The mission called for Lencioni's and Flaherty's platoons to parallel each other, straddling both sides of the river to flush out the enemy. Colonel John H. Cushman (commander of the 2nd Brigade) planned the operation along with Captain Guy Holland III. Colonel Cushman always kept it simple, operating by three governing principles: Work closely with your Vietnamese allies, maintain unrelenting pressure on the enemy, and at every opportunity surround and destroy the enemy.

Cushman's Screaming Eagle Airborne brigade was getting quite a reputation among the enemy in the north. Captured NVA prisoners, while being interrogated, quoted their company commander's orders thus: "If you make contact with the airborne soldiers, get out fast. They will surround and kill you." Another captured NVA prisoner pointed to the Screaming Eagle patch on a nearby soldier, saying, "That little bird is real mean."

Captain Holland had the reputation among his men as being the hardest of the hard-core officers and in Vietnam that says a lot. Captain Holland was on his third tour of duty when he became the company commander of Delta company 1/501. During that tour Holland was hit by the friendly fire of a 105 high explosive Howitzer round that landed short. A golf ball sized piece of shrapnel lodged in his chest creating a sucking chest wound. The injury was so severe that it actually put him into a coma. Within a year hard-core Holland recovered and returned to combat serving his 4th and 5th tours in Vietnam.

Holland had zero patience and or respect for the fresh out of school Lieutenants that were shipped to Nam. He wouldn't even dignify them by calling them Lieutenants instead he referred to them as "Louie's" or any other profanity laced name he could think of at the time. Whether he hated Flaherty more than the other Lieutenants would depend on who you asked but to many it seemed that Flaherty's platoon was sent on more "suicide missions" than any other platoon. Some would say that Flaherty was forged by fire by surviving Holland's hard-core missions and that's what taught Flaherty to be an aggressive but highly competent leader. Others however would argue that Holland was simply just trying to rid himself of Flaherty by either getting him killed or wounded so a more experienced Lieutenant could take his place. Sending Flaherty and Lencioni back into the Eight click Ville certainly supported that second argument

Inside the dense jungle foliage Flaherty checks his map one last time, estimating his platoon is now only two klicks north of the Eight Klick Ville. Even in these early morning hours the heat and humidity are oppressive. The men not only have to fight the enemy; they also have to deal with malaria-carrying hummingbird-sized mosquitoes and hungry leeches.

Out in the field soldiers weren't supposed to use their mosquito repellent (AKA "Bug Juice") because its strong smell would give away their position to the enemy. So, in the bush they have to endure non-stop aerial assaults. At least the leeches are polite enough to bite gently into their hosts' flesh while taking their pint of blood.

Flaherty keeps on top of his men, checking them for leeches, making sure they drink plenty of water and take their giant-sized malaria tablets. In the bush he also observes them for the early warning signs of heat stroke.

As Flaherty's platoon enters deeper into the jungle one of his men near the front element spots movement. The soldier hand signals for the platoon to halt; his signal is quickly passed up and down the long line. The men space themselves approximately ten yards apart — all take a knee to search for targets.

Flaherty — walking near the platoon's middle — starts to carefully work his way toward the soldier who spotted movement. The faint sound of shots echoes in the distance; to the experienced soldiers the gunfire is clear as a conversation. AK-47s ask questions and M-16s answer them. Firefights usually don't last longer than thirty seconds because of the enemy's hit-and-run tactics. However, when they choose to slug it out the gunfire exchanges go from a quick conversation to a long and heated argument.

Before Flaherty can get over to the soldier two Vietnamese men, dressed in plain gray garb, decide it's time to make their move. Concealed one hundred yards away in the thick brush, they first saw the platoon earlier as it crossed an open rice paddy before entering the tangle of jungle, turning in their direction.

Both Vietnamese men take off running low to the ground, away from Flaherty's platoon. One of the men is carrying a long rifle-like object in his hands.

An American soldier points his rifle towards the fleeing men, but at such distance he can't tell if the object is an AK-47 or a gardening tool. Flaherty arrives just as the two fleeing Vietnamese vanish out of sight. Flaherty looks at the soldier and asks, "Why didn't you get a shot off?"

"LT, they took off so fast ... I couldn't tell if they were VC."

Flaherty's jaw clenches as he replies, "Goddamnit, this is a free-fire zone."

The radio-telephone operator (or RTO) is the man responsible for carrying the platoon's heavy PRC-25 radio, better known as the "prick" due to its twenty-five-pound weight and the fact it makes the carrier a higher-valued enemy target. Flaherty's RTO now jogs over to him, announcing, "Delta One-Six is raising you."

Flaherty quickly walks to the RTO man, snatching up the hand receiver (which looks like an old telephone handset).

"Delta Juliet-Six, go ahead," Flaherty says into the handset.

Directly across the river Lencioni is kneeling next to his RTO, holding the hand receiver. His platoon just made contact with several black-pajama-clad VC, resulting in a brief firefight. No one in Lencioni's platoon was hurt, but they killed one enemy combatant and seriously wounded two others. The wounded VC were dragged back into a makeshift perimeter set in a semi-circle, with the river covering their backs.

Lencioni says into the hand receiver, "One-Six, we engaged some light resistance and have two enemy W-I-

As. We are going to call for extraction of the prisoners. Juliet-Six, stand by until the bird picks them up. Over."

Flaherty thinks for a second and replies, "One-Six, roger that. Understand you have two K-I-As?"

Lencioni smoothly replies, "That's a negative. I have two wounded, I repeat two enemy W-I-As."

Flaherty's hand tightly clenches the receiver as he repeats, "One-Six, verify again. You advised two K-I-As?"

Lencioni mutters under his breath, "God damn it. I'm not killing them."

He replies into the radio, "Negative. Negative, Juliet-Six. They are W-I-As, I repeat Whiskey India Alphas, and we're waiting for helicopter extraction. Hold your position. One-Six, Wilco, out."

Lencioni angrily throws the handset back to his RTO man. Flaherty frustratedly hands the set back to his.

The RTO turns to Flaherty and asks, "LT, I don't understand. Why did you ... "

Flaherty looks around at the terrain, not liking what he sees. *With the recent gunfire every VC and NVA in the area will know exactly where we are. Even if they don't jump us the enemy mortars will be raining down on us soon.*

He cuts off his RTO man, saying, "This platoon doesn't hesitate to fire in a free-fire zone, and this platoon doesn't have the luxury of taking prisoners. You hesitate out here and you'll be flying home in a bag. Now get back on the radio and advise 2nd Platoon that we spotted two suspect VC, last seen heading their way."

"Yes, sir."

Flaherty calls his sergeant over and angrily says, "Pass it along: we are not going to leave 2nd Platoon. We're going to sit tight with our asses exposed while they wait for an extraction."

Landing Zone Sally exemplified the extremes of the country it sat in. During the rainy season the camp flooded into a muddy mosquito-laden bog, and during the dry season it was an arid desert-like dust bowl. This was the third week without rain and fine layers of dirt were constantly being kicked into the air, making hazy reddish clouds.

Lieutenant Lencioni sits outside his hooch on a small wooden chair, reading after-action reports as the incoming supply trucks race into camp. After each truck passes a new cloud of dust rolls over Lencioni, coating him with a fresh layer of grime.

As the next wave of dust heads towards Lencioni a lone figure emerges from its red cloak. As the figure draws closer Lencioni recognizes the unmistakable shape of Richard Flaherty. Flaherty is strolling loosely towards him, drinking from a small metal flask. He isn't carrying any equipment. Buckled about his waist is a duty belt holding a pearl-handled Colt Python revolver — cowboy-style in a low-slung leather holster.

Flaherty approaches Lencioni with a determined grin on his face, tucking the flask into his cargo pants pocket. Lencioni looks up just in time to see Flaherty fast draw the gun and aim it straight at him. Flaherty then — with the dexterity of a Hollywood cowboy — spins the gun on his finger and snaps it back into his holster.

Lencioni, without taking his eyes off of Flaherty, slowly starts to place the reports on the ground. Flaherty quickly repeats his quick draw action, ending it once again with the revolver pointing straight at Lencioni.

Lencioni — still seated — slowly reaches for his M-16, which is leaning up against a wall next to the chair. The resolve on Flaherty's face gets stiffer as he quickly re-holsters his gun.

Lencioni, both hands now on his rifle, smoothly and unhurriedly swings the weapon in front of him. *I don't know what's pissing me off more, him pointing that revolver at me or the arrogant grin on his face.*

He uses his thumb to switch the rifle from semi to fully automatic. It makes an audible click, which Flaherty acknowledges with a slight nod. Lencioni slowly raises his rifle's barrel, pointing it toward Flaherty.

"Richard, if you draw that gun one more time I'm going to stitch you from your head down to your fucking toes."

Flaherty welcomes the challenge, steadying his right hand as it hovers over his holstered gun.

Lencioni leans the butt of his rifle deeper into his shoulder. The two alpha males lock eyes, searching each other's souls for any sign of weakness.

Flaherty reaches down and slowly rests his hand on the grip of his gun. Lencioni closes one eye and slides the pad of his index finger onto the trigger, starting the slow squeeze. 7.9 pounds of force on the deep notch pull was all Lencioni needed to fire his M-16, and he was already close to exerting five pounds.

Flaherty lifts his left hand into the air palm forward in a sign of surrender. Slowly he pulls the revolver out of its holster, keeping the barrel pointed to the ground. With his thumb he moves the cylinder release latch, dropping open the cylinder. He leans the open gun forward so Lencioni can see it has no bullets in it.

Flaherty snaps the cylinder shut and replaces the gun in its holster. He casually says to Lencioni as he strolls by, "Foul times require foul tests. I needed to see what you were made of."

Lencioni watches as Flaherty walks deeper into camp. He waits until he's completely out of sight before calmly putting down his rifle. He once again casually picks up his reports, muttering out loud to himself, "Strange little motherfucker."

12

The Silver Star

April 19, 1968 — 0630 Hours
LZ SALLY
TOC Command Bunker

Colonel John H. Cushman, forty-seven years old with the stalking presence of a born leader, stands in front of a map addressing Captain Holland and a large group of his airborne officers. Inside the bunker, Flaherty stands next to Lencioni. The two men have worked several missions together since the gun incident, and neither one has said a word about it. In Vietnam such exchanges quickly pale to insignificance.

Cushman continues his briefing: "With the successful ending of Operation Carentan, we will now be commencing with Carentan II. However, your units were specifically chosen to conduct Operation Delaware. Your primary mission will be to block enemy supply routes exiting the A Shau Valley as they head towards Hue."

Cushman points to several marked areas of his map and adds, "We'll be setting up blocking forces here off to the east with First and Fourth Platoons. Lieutenant Lencioni with Second Platoon and my one-meter Lieutenant Mr. Flaherty with Third will link up and support those platoons as you move toward the A Shau Valley."

An army platoon resembling a giant green boa constrictor stealthily snakes its way in a single file line through thick jungle. The terrain that morning dictated the formation however on most patrols Flaherty preferred utilizing an arrowhead formation with a left and right flank. The head of the serpent is led by the point man; being asked to walk point can be the highest honor or worst punishment, depending on who you ask. His responsibility is to walk headlong into hostile territory, searching for the enemy and his traps.

Drenched in sweat, the point man (Mike Nunez) keeps his head on a swivel, scanning every inch of jungle in front of him. His nose picks up the slightest wisp of a foreign scent wafting above the rotting vegetation, and his body automatically freezes. The rest of the serpent comes to a halt, coiled and ready to strike. Walking behind the point man is the slackman, whose responsibility is to check for anything the point man might have missed.

Nunez turns to the slackman and whispers, "Fitz, up ahead." Fitz scans the jungle, looking for movement or any shape or color that doesn't blend in.

"I don't see anything," Fitz whispers back.

Nunez nods his head, indicating the way forward. "Someone was just smoking a cigarette ... "

The jungle erupts into a fiery orange explosion of AK-47 and rocket-propelled grenade fire. An RPG round detonates on a tree just feet from Nunez, instantly shredding him to pieces. The line falls back as the incoming fire increases tenfold.

RPG rockets whoosh through the air, cutting through swaths of jungle. The rockets impact onto the larger trees, causing thunderous explosions that rain shrapnel and debris onto the platoon as they hunker deeper into the muddy jungle floor.

Flaherty runs forward as bullets whiz by, diving behind a downed tree next to Fitz.

"Where's Nunez?"

"He's, he's ... " Fitz is trying to process what just happened to his good friend.

Flaherty, trying to snap him out of it, interrupts, "C'mon Fitz, get your head in the game."

"He's everywhere. They blew him to bits."

"Okay Fitzy, you're the spear tip now. No one gets by you. If we don't start flanking them they'll mow us all down."

The incoming fire continues to probe their position. Flaherty removes three grenades from his vest and puts them down next to Fitz.

Flaherty roughly grabs Fitz's arm, making sure he has his full attention, and adds: "Every thirty seconds lob one, and conserve your ammo to short bursts." Fitz nods his head.

Flaherty — in a low crawl — heads back towards the rest of his men. As he continues to edge toward his RTO two rocket-propelled grenades whiz over his head. KABOOM! A third RPG explodes against a tree just in front of Flaherty, lifting him airborne and spinning him like a top.

As his helmet flies off his head Flaherty is momentarily knocked unconscious. He awakens to sharp pain: a sliver of metal shrapnel is burning into the top of his skull, causing blood to flow freely down his face.

RTO Orlando Lewis grabs Flaherty by the sleeve and drags him behind a log. Lewis sees the small piece of shrapnel and gently pulls it out of Flaherty's head.

"LT, are you okay?"

Flaherty is experiencing blurred vision and a loud buzzing sound in his ears. Lewis' voice is muffled, distant.

"What?" Flaherty asks, not even able to hear his own voice.

"LT, are you alright?"

No response.

"Medic! I need a medic!" Lewis screams.

The buzzing starts to subside, and Flaherty's eyes begin to refocus.

As the sounds of violent assault start to flood his ears again, Flaherty answers, "I'm fine. I'm fine. No medic. Did you call in the contact?"

"Done. They're working on an artillery firing solution." Lewis ducks into the dirt as another RPG whooshes over them.

"Good job. Until that arty starts coming in we need to put some pressure on their right flank. I think we hit a fortified bunker."

Flaherty leaves Lewis and runs over to Sergeant Hawkins. Bullets like angry wasps follow him, trying to exact their revenge. "Where's my arty?" he barks over his shoulder. "I need that bunker taken out!"

Hawkins buries his head momentarily into the dirt as several AK-47 rounds zing past, then answers, "First and Second Platoon are getting clobbered. They're using all of the artillery to keep from being overrun."

Flaherty fires a volley of M-16 rounds, turns to Hawkins and says, "Shit. Okay, sit tight."

Flaherty crawls towards several of his soldiers holding the front line. He motions to three of the men, positioning himself behind a large downed tree.

"Griggs, Escobar, Brody, fall back to my position."

The three soldiers nod. Flaherty slaps in a fresh magazine and shouts, "Cover fire!" while letting loose a long volley.

The three young soldiers move faster than they ever crawled in boot camp, making record time slinking their way over to Flaherty.

Flaherty stops firing and turns to the men. "We have no fire support, and that bunker is chewing us up." He points to a raised berm twenty yards forward. "I'm going to advance to that berm. You see it?"

All three men see the berm. They nod and whisper collectively," Yes sir."

"Jimmy, I want you to grab O'Meara's 90-millimeter rifle. Do you know how to fire it?"

Jimmy Brody nods yes.

"Okay, good man. O'Meara's back there getting his leg patched up — his fire team is KIA. Once I draw their fire you get in position by that tree stump and let go with a round every ten seconds. After a couple rounds make sure to let it cool for a few minutes or you'll cook the rounds off in the tube. Grab Sikes and have him carry the ammo and load for you."

Brody nods and says, "Got it." He turns on his heel and sprints towards the inner perimeter.

Flaherty continues with, "Guys, as Jimmy's launching rounds at the bunker I want you to flank them to the east. Head off to that bomb crater and wait 'til you see me make my charge."

Griggs and Escobar stealthily slide on their stomachs over thirty yards to the east and wait for Flaherty. Brody runs up to Flaherty, shouldering a large and heavy recoilless rifle that looks similar to a WWII Bazooka. Sikes, his loader, comes behind him holding a big satchel of ammo.

"Jimmy, if you get a direct hit watch out for Griggs and Escobar. They'll roll up to the bunker if they get the opportunity. Ready?"

"Fuckin' A."

"Okay boys, here we go."

Flaherty takes off in a sprint, heading just west of the bunker. Immediately all the machine gun fire from the bunker is aimed towards him. Bullets chew up the ground near Flaherty's quick-moving feet; a bullet strikes the heel of his boot, knocking him to the ground. Flaherty quickly scrambles upright, limping as he sprints the last ten yards to the treeline.

Brody shoulders the large rifle tube. With Sikes on his heels carrying the ammo he runs straight up the middle, hunkering down behind the huge tree stump. In a heavy sweat, Flaherty makes it behind a tree and immediately starts firing three-round bursts at the bunker. All enemy fire is now concentrated on Flaherty.

The loud boom! of the recoilless rifle is immediately followed by its round exploding near the enemy bunker. The explosion doesn't knock out the bunker, but launches a large cloud of dirt and debris into the air.

Griggs and Brody sprint towards the east side of the bunker. The tree Flaherty is using as cover is getting shredded by the withering machine gun fire. Debris from the tree is knocked into Flaherty's eyes; he's temporarily blinded.

Griggs and Escobar watch the bunker, waiting for their opportunity to launch their attack. Brody fires off another round — a direct hit to the bunker's front-facing sandbag wall. Griggs and Escobar use the opportunity to take off in a full sprint towards the smoldering bunker, firing from the hip as they run. They kill several now-stunned NVA manning the machine gun inside the bunker.

Two other NVA soldiers try to sneak away by crawling through a rear trench. Griggs shoots the legs of the first man, and the other immediately throws down his AK-47 and surrenders.

Flaherty finally clears the debris from his red and swollen eyes. Looking up, he sees Griggs and Escobar walking behind two NVA soldiers. The prisoners have their hands up. One of the NVA is wounded badly and is having trouble walking.

Griggs says over his shoulder to Escobar, "Cover him, and I'll drag this dink back."

Griggs hands his rifle to Escobar and walks to the wounded man. The wounded NVA is bent over in pain, blood flowing down his legs.

"Nooo!" Flaherty screams in warning.

As Griggs tries to grab the wounded man's shirt the man pulls a small knife from his boot, plunging it deep into Griggs' stomach.

Running towards the men, Flaherty expertly places a

well grouped three-round burst into the knife-wielding NVA's chest, instantly dropping him. The other NVA soldier turns his body towards Flaherty, who without hesitation drops the man with one clean shot to the head. Escobar grabs Griggs, who is holding his stomach, trying to stop the flow of blood.

Flaherty barks, "Get him back to the doc." Then he shouts to all the men in his vicinity:

"From now on until I say different ... we don't take prisoners!"

As Flaherty and his three men work their way back towards their defensive perimeter, hundreds of NVA sprint through the jungle towards them.

April 19, 1968 — 1445 Hours
LZ SALLY
TOC Command Bunker

Colonel Cushman, Captain Holland, and several other officers are crowded around a radio. Lt. Lencioni, covered in dirt and blood, runs into the tent and asks Holland, "Sitrep, sir?"

Holland understanding Lencioni's grave concern ignores the flagrant violation of a junior officer asking a senior officer for a "Sitrep" and shakes his head. "You and First Platoon were the last to make it out before the trap was sprung. Flaherty's platoon and Fourth platoon are completely surrounded. The last report is three KIAs and fifteen WIAs."

"Any gunships in the area?"

"That's a negative, lieutenant. Almost all air support is tasked for Operation Carentan, and already several units have engaged stiff resistance in Dong Ha."

Holland walks over to the war map and draws a circle surrounded by Xs. He starts explaining to Lencioni, bringing him up to speed. "Fourth platoon leader is KIA.

Flaherty was able to fight his way over to them and consolidate the two platoons. He's just finishing setting up the defensive perimeter. We're working on logistics for a resupply airdrop."

Lencioni shakes his head, saying, "That perimeter is too small, and the jungle canopy is too thick for an airdrop. We'd be just giving them away to the enemy."

April 19, 1968 — 1500 Hours
Thua Thien Province

Flaherty is limping back and forth, placing men into different firing positions. Constant AK-47 fire pours in from all directions, making it even more difficult to move the men. After Flaherty finishes his placements, he hobbles back towards their makeshift triage area for the wounded. More than fifteen of his men are either unconscious or heavily bleeding. One medic is feverishly working on the injured. *Time is running out — we gotta get these men outta here!*

April 19, 1968 — 1615 Hours
TOC Command Bunker

Cushman turns to Holland. "The LZ's too hot for an extraction. They're going to have to gut it out 'til morning."

"They won't last the next few hours out there," Holland replies. Lencioni turns to Holland points at the map and adds, "Have the choppers insert my platoon back onto this ridge. We'll hump and fight our way over to them."

"That's a no go and even if we had air support that would be a suicide mission. I can't risk losing anymore men." Lencioni asks, "Why aren't they working the arty?"

"A small group possibly a squad of men from Fourth

got trapped somewhere outside the perimeter," answers another captain.

Lencioni stares at the map in frustration. "Damn it."

April 19, 1968 — 1830 Hours
Thua Thien Province

Several of Flaherty's men are working to set up claymore mines and trip wires just outside the perimeter. Inside the perimeter, the men are making last-minute efforts to dig in and create some type of fighting holes. By the time they're finished the last of the day's light is gone. Night, with its unseen terrors, has come.

Flaherty is talking into the radio handset. "Juliet-Six I copy. We'll hold tight." He hangs up the receiver and runs to his makeshift triage location. Surrounded by large downed trees in a natural depression, it's the safest area inside the perimeter. Flaherty checks with his medic, who's still frantically working on recently hit Sergeant King. King was a salty lifer who had served in Korea and was on his third tour in Vietnam. He was hit after the third time he valiantly rushed outside the perimeter to retrieve fallen men. King going down was an enormous blow to the platoon's morale. Most believed he was the man that would get them all home safe.

After hours of constant incoming fire the intensity of the enemy barrage starts to ebb, with only sporadic crackling from the AK-47s. Flaherty looks around, sensing something is wrong. What he doesn't see are six Viet Cong, dressed in their black pajamas, stealthily snaking their way on their stomachs towards Flaherty's perimeter. The NVA expertly locate and cut the claymore mines and trip flare wires left on the perimeter by the Americans. Over a hundred VC and NVA soldiers silently start inching their way forward.

Flaherty goes over to his main defensive line, the bulk of his platoon positioned here based on the area's topography.

He starts to address the men. "They're about to hit us hard again — I can feel it. I want everyone ... "

Flaherty hears screaming behind him. He turns to see three of his men struggling to hold down another soldier.

Flaherty rushes over to the commotion. "What the hell is going on?"

The frantic soldier is Joseph Cummings. He stops struggling when he sees Flaherty approach. The M60 gunner Louie Fernandez (who is straddling Cummings) gets off him, but still grips him tightly by his uniform jacket.

Fernandez turns to Flaherty. "Cummings flipped out after King went down. He's not making any sense, and he says we gotta surrender." The other two soldiers loosen their grip on Cummings's limbs as Cummings's breathing starts to slow.

Flaherty takes off his helmet and kneels in front of Cummings. "Joe, I need every man that can fight on that line. There is no surrender. You got that?" Cummings stares blankly at Flaherty.

Flaherty takes a few breaths and calmly adds, "Son, I need you to hold it together. When we're out of the bush I will personally make sure to transfer you or send you home. But for now you will grab your rifle and you will fight, or I will personally tie you to that tree." Flaherty points to a half-knocked-down palm tree that has absorbed numerous RPG rounds.

Cummings looks around as he starts to register where he is. Vaguely, he nods his head yes to Flaherty.

Flaherty slaps him on the shoulder and says, "Good man." He turns to Fernandez. "Give him back his rifle." Flaherty looks around at his men, seeing that Cummings's outburst has further hurt morale. A nervous look of uneasiness starts spreading like an infection. Flaherty knows how quickly fear can turn into panic.

Flaherty stands up, exposing himself to danger but showing the men he is unafraid. In a loud assuring voice he says, "Listen up. The enemy is preparing a full-out frontal assault, hoping to overwhelm us with numbers. Hoping we'll break and run. There is no place to run — and they're not here to take prisoners. We make our stand right here and right now."

Flaherty pauses, and the platoon focuses on what he will say next.

"I promise if you do exactly what I say and rely on your training, Charlie will cut and run. We'll all make it out of here safe and sound. Now, focus on your targets, make every bullet count, and make Charlie pay for every fucking inch 'til he can't take it no more!"

A look of determination floods over the men's faces as they look at each other, acknowledging their brotherhood. Just then a soldier spots movement and fires his rifle on fully automatic. The NVA are now on their feet in a light jog, screaming their battle cry as they advance. Flaherty bends to a knee and starts firing his rifle at the numerous targets. All his men join him in a deafening symphony of death.

Cummings uses the opportunity to grab a white handkerchief he had hidden in his pocket. He attaches it to his rifle, then waves the modified white flag over his head as he runs toward the enemy.

Flaherty sees the man run and yells, "Cummings, get back here you son of a bitch!"

Two of the soldiers to Flaherty's right stop firing. Cummings is inside their field of fire — they're afraid they might hit him. Cummings dives to the ground as rounds fly over his head.

Flaherty continues to knock down attackers with quick three-round bursts. He screams over the battle noise to the men next to him, "If he gets up again you do not stop firing. If he gets hit ... then he gets hit. That's an order!"

Despite their losses, the huge enemy wave continues

to crash forward. Cummings finally comes to his senses and crawls back to the platoon unharmed. The first wave of charging enemy soldiers finally stalls only twenty yards from Flaherty's line. An enemy commander, using his whistle, signals the retreat; they pull back with as many of their fallen comrades as they can grab.

April 19, 1968 — 2130 Hours
TOC Command Bunker

Inside Colonel Cushman's command bunker men are huddled around the radio, listening to sounds of the horrific battle going on. Cushman reads through reports and continues to issue orders. He turns to Holland and encouragingly says,

"They're still hanging in there."

"But for how much longer?" Lencioni says out loud. "They need to find those lost men."

April 20, 1968 — 0130 Hours
Thua Thien Province

As the enemy retreats Flaherty shouts, "Cease, fire. Cease, fire." The ceasefire order is passed along, shouted from man to man; gradually the gunfire dies down. Sounds of battle are replaced by screams of agony from the wounded and dying.

Flaherty turns to Sergeant Rob Kane. "We ain't going to survive the next charge without artillery support. Send a couple men out to check our claymore and trip flare wires, 'cause they should have gone off in that last attack."

"Yes sir," Kane responds.

"Next, tell them to grab the smallest dead NVA they can find and drag him back here with all his weapons and equipment."

"LT, you want a dead gook?"

"You heard me. Just do it."

Flaherty strips off his clothes as two soldiers drag a small dead NVA soldier back into the perimeter.

Kane looks at Flaherty, shaking his head in disbelief. "You can't be serious."

Flaherty starts removing the dead soldier's uniform. "It's the one time my height will actually help."

Flaherty — wearing the NVA uniform — places the NVA helmet on his head, partially obscuring his face. Kane walks over and smears mud across Flaherty's face, helping to further disguise him. Flaherty picks up the dead man's AK-47 rifle and crawls out of the perimeter towards the enemy.

Ten yards out Flaherty sees several dead NVA soldiers and grabs the nearest one. He slings the AK-47 over his shoulder and drags the dead soldier backwards by his uniform collar, heading toward the enemy lines.

Looking over his shoulder he sees several pockets of NVA soldiers wave to him, motioning him to keep dragging the man back past them. After Flaherty passes the first line of enemy lookouts he drops the body and starts scanning for his lost men.

A group of NVA are moving forward towards Flaherty — he's caught in the open. *Shit!* Flaherty quickly kneels and starts disassembling his AK-47 on the ground.

The soldiers walk past Flaherty. The last NVA to pass him by is an officer. "*Tro lai,*" he barks.

Flaherty nods, keeping his head down as he starts to reassemble the rifle. The soldiers continue to move forward towards the besieged American platoons.

Flaherty scans the area in the darkness, making out the shape of several previously destroyed enemy bunkers. As he approaches each he utters in a hushed voice the password:

"Purple haze. Purple haze."

Inside the bunker a dead NVA soldier starts to move.

From beneath the body a U.S. soldier cautiously peers out and whispers, "Goofy grape."

Flaherty — already starting to scurry away to the next bunker — stops as he barely hears the return password.

Flaherty turns back to the bunker and slowly pulls out his Buck 119 hunting knife, whispering, "Purple haze?"

"Goofy grape," replies the soldier again.

Flaherty quickly returns his knife to its sheath and jumps into the bunker. He holds a finger up to his lips, warning the men inside to be quiet. Three soldiers push dead bodies aside and scramble towards Flaherty.

One of the soldiers, with blood streaming out of his ears, loudly says, "LT we got ... "

Flaherty motions sharply with his hands as he says, "Keep it down! There's enemy less than fifty yards away. Where's the rest of the guys?"

The first soldier says, "Follow me."

The first soldier crawls out of the bunker, Flaherty and the other two men following. He leads the group twenty yards away towards an old deep artillery crater full of scattered debris. As he gets closer to the crater he starts whispering, "Purple haze. Purple haze."

Hidden under dirt, leaves, and palm fronds the last three lost soldiers slowly rise from their hidden graves.

"Is there anyone else?" Flaherty asks the new group. The soldiers shake their heads no.

"Stay on my ass. If they spot us ... make a run for it." He points in a forward direction and adds, "The platoon is about one hundred and fifty yards over there."

The seven men cautiously work their way through the jungle. Just fifty yards from their platoon they are spotted by the NVA. One of the American soldiers is shot in the back and collapses.

"GO! GO!" shouts Flaherty. He tries to grab his fallen comrade and is nicked in the shoulder by a bullet.

One of the other soldiers tries to grab the downed man, but Flaherty waves him off. "He's dead. Run."

The first soldier in line running towards the platoon waves his arms about frantically, yelling, "Purple Haze! Purple Haze!"

Several U.S. soldiers break from their perimeter, running out to help Flaherty and the five soldiers. The NVA use the opportunity to launch another large-scale attack.

As soon as all the men are safely inside the perimeter Flaherty shouts to Lewis, "Start raining down that fucking artillery!"

The friendly thunder of giant 155 Howitzers — coming from Phu Bai — starts booming. Artillery shells rain down onto the advancing NVA, sweeping them aside like an angry hand clearing a chessboard.

April 20, 1968 — 0630 Hours
Thua Thien Province

Five Huey helicopters packed with soldiers fly in formation as the sun rises from the horizon. The lead chopper carries Colonel Cushman, Lencioni, and several other officers. The convoy flies over Flaherty's still-smoking perimeter.

The fighting is over, and their birds-eye view reveals the extent of the carnage. Lencioni sees close to a hundred dead enemy soldiers strewn in concentric rings around the perimeter. Mixed in the rings of NVA soldiers are body parts and shredded web gear, completing the gruesome display.

Following protocol, Lencioni requests over the radio for someone in Flaherty's unit to 'pop smoke' before they set their choppers down. One of Flaherty's men opens a smoke canister containing a bluish smoke; it quickly floats over the perimeter. Lencioni responds, "That's affirmative. I have bikini blue." Lencioni looks down on the battlefield, this time seeing a majestic vision.

The bluish smoke drifts in between long red-and-white

strips of bloody gauze flapping off the bushes, while the sun's rays reflect and glimmer off a mound of empty c-ration and expended ammo cans. It completes the most authentic American flag he's ever laid eyes on.

Cushman and Lencioni, along with a fresh platoon of soldiers, enter Flaherty's perimeter. The churned-up earth looks more like the dark side of the moon than a tropical jungle. Huge craters and baseball-sized rocks litter the area.

Flaherty is sitting on an ammo box as a medic patches up his shoulder. He and his men look haggard, covered in blood and dirt. Almost every man is in a reclining position or sitting quietly in reflection; some are smoking cigarettes. The wounded soldiers quickly receive attention from a group of medics. Once stabilized they're placed on litters and loaded into medevac choppers, known as dust-offs, for a quick hop to the nearest medical units for more attention.

Flaherty sees Lencioni first and nonchalantly says, "I hope you didn't forget the coffee and doughnuts."

Lencioni smiles and replies, "Hardnose little bastard." Lencioni puts his arm around Flaherty and pats him on the back a couple of times. Cushman walks over to Flaherty, and he quickly stands at attention.

Cushman proudly says, "At ease, son. I want you to know I'm putting you in for the Silver Star."

"Not me, sir. My men are the ones that deserve all . . ."

"Save the humility shit for the AP reporters, 'cause the Airborne needs officers like you. I'm also having you promoted to First Lieutenant."

Flaherty, without any fear, counters, "I appreciate that, sir, but after this tour I was hoping to get a shot at Special Forces school."

"Well, I'll be a monkey's uncle. My one-meter lieutenant wants to be a Green Beret." Cushman frowns for a moment, then shrugs. "I could try to talk you out of it, but I know what a stubborn little prick you are."

"Thank you, sir," Flaherty proudly answers.

For his heroism on April 20, 1968, Lieutenant Richard Flaherty was presented with the Silver Star. It was just one of many medals, awards, and commendations he would receive.

On May 17, 1968, Operation Carentan II came to an end. One hundred and forty U.S. soldiers would be listed as Killed in Action, seven hundred thirty-one listed as Wounded in Action, and forty-seven declared Missing. Enemy Killed in Action were eight hundred sixty-nine, with an undetermined amount for their Wounded in Action.

Ironically, I later learned that the battleship USS New Jersey delivered much of the firepower and shelling in the Quang Tri region where Richard's platoon operated. The New Jersey was the same battleship his uncle Joseph Ambrose Flaherty served on during World War II, twenty-five years earlier; he lost most of his hearing due to the concussions of the main guns firing.

13

Snakes and Rats

Kosher Kingdom Bench

While downing my Gatorade I asked Richard, "Were you ever afraid of getting killed over there?"

"No, not really. I just always had this feeling that I wasn't going to die over there. Although there was this one time.... I came real close to buying the farm when I almost drowned."

"River crossing?" I guessed.

"No. In a rice paddy."

"I thought the water was only a foot or two deep in a rice paddy?"

"Yeah, usually it is. But this was during a typhoon — I believe it was Typhoon Bess."

1968 — LZ Mongoose

"I was in charge of RECON Platoon, Echo Company, I think around September of '68, and we were operating out of LZ Mongoose. We were out in the field setting up an ambush; it was already raining pretty hard when we got hit with these huge bands of rain from a typhoon. And I'm not talking heavy rain: I'm literally talking about buckets of water pouring down on us. It must have been high tide, because the storm surge pushed the ocean miles inshore to where we were operating.

"The rice paddy couldn't have been more than one hundred and fifty yards long, but once that typhoon hit it took us hours to cross it. Recon was a small unit, so we'd bring as much ammo as we could carry; sometimes we'd be humping over seventy pounds of ammo and equipment on our backs.

"The water in that paddy must have been over five feet high, because I was completely underwater once it hit. One minute I was slogging in waist-high water, the next I was trapped underwater for at least thirty seconds. I couldn't get the equipment off of me fast enough — everything was just muddy and black. One of my guys must have seen me go under, reached down and pulled me up. Otherwise I would have drowned. I was for sure a goner.

"Two of my guys, Al Dove and Jerry Austin, had to take turns carrying me on their backs for most of that walk 'til we made it onto this tiny little island. I also remember that night because Al Dove shot some NVA who were also trying to save themselves and get up on our little island. One of those soldiers was carrying all these plans and maps detailing an upcoming sneak attack on one of our bases. It was really a big score to recover that much intelligence, and it definitely saved a lot of lives.

"After killing the NVA we got into another big firefight, but this time it was with snakes and rats."

I turned to Richard. "I'm sorry ... Snakes and Rats? Is that some code name for the enemy?"

"No, I mean real snakes and rats. 'Cause of all the flooding, they were trying to get onto our little island, swarming out of their holes. These weren't your regular type of rats: these things were huge, the size of small dogs. And most of the snakes in Vietnam were poisonous; I mean vipers and cobras. So we were really fighting for our lives that night." "What about being shot? Didn't you constantly worry about that?" I asked.

"To be honest with you, no. I'm not trying to sound

macho or anything like that — I'm just being practical. Most Vietnamese at a distance would probably hesitate to shoot at me, thinking I'm one of their own. They also calibrated their machine guns to a certain height which usually went over my head, and besides everything else the Vietnamese were really bad at shooting.""Bad training?" I asked.

"That's one reason. The other is the weapon they carried, the AK-47. With an automatic rifle, they tend to just put a lot of rounds down range — spray and pray. They don't really aim that much.

"Now we do it differently. American military training, ever since the Revolutionary War, and even more so in this last century, is all about marksmanship. We're really heavily graded on marksmanship throughout our training. Not all armies do that. Especially guerrilla armies, militias and the like. When we shoot, we tend to hit what we shoot at. We don't try to put a whole lot of random rounds down range."Richard stopped for a moment, lost in a memory. He started laughing.

"What happened?" I asked, smiling.

"I almost forgot about this one time. I guess talking about how bad the Vietnamese were at shooting reminded me of a story of how good this one guy must've been.

"Every once in a while we'd have a day or two of stand-down time, so we would head over to CoCo Beach, which was really close to our LZ.

"Now, it was almost all level ground for miles around that beach. It was a safe area as far as we were concerned, and it was a nice-looking beach. At the time I was with the same group of guys from my Recon unit, and we were just swimming and sunning ourselves — which was a good thing, 'cause the salt water in the ocean also cured our jungle rot.

"So, the guys were laying out on the beach and I went for a swim. I rolled over to my back and started to do this lazy backstroke, the water just as calm as it could be. I'm looking up at the sky and my ears are under water, but I start hearing this school of fish jumping all around me.

"I think nothing of it until one of these fish jumps so close I feel a splash of water on my face. When I look up I see my guys on their bellies waving at me, hollering, 'Sniper!' I ducked under the water and must have made it back to the beach in Olympic record time. By the time I reached shore all the shooting had stopped, and the shooter must have beat feet.

"I can't even imagine how far that sniper must have been from us. He had to be close to half a mile away. The fact he was able to group his shots so close to me makes him one hell of a sniper in my book."

———————

Only days before I had to hand my final draft of this book to my editor for the release, I was contacted by Warren W. Chan, George Page and Norm Campbell. All three men served in the 101st Airborne with Flaherty in Vietnam. Warren Chan wrote me the following:

"My fondest memory of Lieutenant Flaherty was during the time we were both stationed at a base in Cu Chi [southern part of Vietnam]. Delta Company had just moved up north to I Corps, [Quan Tri district] and we were left behind to clean up some loose ends before we followed them up north.

"After finishing up our duties for the day it was movie time, so I grabbed my case of 3.2 beer under one arm and headed to the outdoor screen where they showed the movies. I sat next to Flaherty and without even look-ing he reached out his tiny open hand which I quickly filled with a beer. That night Flaherty was the officer in charge, so he probably shouldn't have been drinking, but he seemed like he already had a couple of belts of bac si de [home-brewed rice whiskey] in him. The movie they were showing was "Bullitt" starring Steve McQueen. The movie starts and I'm just happily drinking away when all of a sudden BOOM... BOOM... our camp starts receiving enemy mortar fire.

"So Flaherty takes off like a bat out of hell and he jumps into this old Korean era U.S. military ambulance (much like ones you would see on the tv show M.A.S.H.). Flaherty, a couple of days earlier, doctored it up so he could drive it. He attached small blocks of wood onto the gas, brake and clutch so he could reach the peddles. Even so, the truck wasn't made for someone Flaherty's size and he couldn't see out the front or side windows. Flaherty starts driving around the camp screaming out the window, 'MORTAR FIRE... MORTAR FIRE...' and all you could see was the top of his tiny head as he drove around the camp in circles. We all thought that was the funniest thing and couldn't stop laughing.

"On another incident, while I was Captain Holland's Radio Telephone Operator, we were all out in the jungle on a search and destroy mission. We were walking through some thick elephant grass, which was this tall razor-edged grass usually about waist-high on a soldier. Captain Holland called Flaherty on the radio, trying to find his location, and Flaherty answers, "I'm right in front of you, Captain." Still not able to see Flaherty, Holland frustratedly tells Flaherty to stand up. Flaherty fierily answered back..."I am standing up." Captain Holland then orders Flaherty to start jumping up and down so that he could see him. Flaherty is now pissed but an order is an order so he starts jumping — but the best he could do is to jump only about a couple of inches cause of weight of gear that he was carrying. We all watched Flaherty angrily bouncing up and down like a pogo stick, waving his arms, and it was really comical."

I spoke by phone with Warren Chan and I was curious about his own unique experience being an Asian-American soldier fighting in Vietnam.

"I only remember meeting one or two other Asian-American soldiers in the Army during that time."

"Did you face any racism because of it?" I asked.

"Maybe, but it was no big deal. Although a bunch of times I do remember, while on missions in the jungle, getting a rifle pointed at me. Usually it would just be the new guys assigned to our company. I would yell out to them, 'I speak better damn English than you do, so put that rifle down before I shove it up your ass!'"

"What about the friction between Captain Holland and Flaherty? I asked.

"Well, Holland, 'the ole' man,' he was as tough as nails and not everyone liked him, but they all respected him. Now did he hate Flaherty more? Maybe, but he seemed to dislike most of the new lieutenants, so at least he was fair about it by treating them all like shit."

After I hung up with Warren, he seemed a little too humble, so I checked up on his military past. Of course, just like most of the men I interviewed that didn't talk about themselves, Warren Chan was a bonafide war hero. His jacket of decorations and medals was a page long, including the Silver Star, 2 Bronze Stars and a Purple Heart with oak leaf cluster.

Warren was also the soldier who first attended to Captain Holland when Holland was hit by the Howitzer short round that caused a sucking chest wound. While applying pressure to stop the flow of blood, Warren radioed in the cease fire order to stop the deadly onslaught of more friendly fire. Two other young men sadly would be killed by those rounds before the shelling stopped. Warren then called for a medevac chopper to extract Holland. Colonel Cushman, who was in his own helicopter in the area, ordered his chopper to land and pick up Holland. Cushman's Chopper quickly got Holland to the nearest hospital, which probably saved his life.

The other man that day in the field next to Warren helping patch up Captain Holland was Sergeant George Page. Warren gave me George's phone number and we ended up talking for almost two hours. George's memories of Flaherty went all the way back to their days in Fort Campbell.

"The first time I met Flaherty was inside the camp and I couldn't believe this tiny guy had such a tough guy attitude. He would challenge anyone by getting in their face and showing them he wasn't a man to be messed with. But the other thing about Flaherty was he was so inspirational he was almost supernatural with his abilities and drive. We would all be tired after running five miles, but Flaherty would keep going and run ten. Everything we did physically, Flaherty would surpass us. I never saw another man with that much determination and grit... never. Well in Nam, he was the same way with that bad-to-the-bone attitude. One of his first leadership tests was during the Tet Offensive, and Flaherty lost a lot of men during that time. Matter of fact our whole company saw its biggest losses during Tet: It really was hell.

"I remember the day Flaherty first got wounded in the leg, because I was also shot. That day I believe we had a total of twenty-two men in our company who were either wounded or killed. That day Captain Holland strapped a radio on my back and sent me and another sergeant to check up on Flaherty and a squad of his men because they weren't answering their radio. Well, Flaherty and his squad were pinned down by heavy fire and their radio was shot to shit. Next thing I know I was shot in the head and knocked unconscious. The bullet ricocheted around inside my helmet, which saved my life, but parts of the bullet entered the back of my skull. I slid unconscious into a rice paddy and would have drowned, but one of my guys pulled me out. When I came to, I was still very woozy, but me and the sergeant raced back to tell Holland about Flaherty's squad being trapped. Holland started yelling at me about why I didn't just radio in the information until I turned around and he saw that I was hit in the back of the head and bleeding. Well, just like hardcore Holland, he sent me right back out into the fight two more times to try to outflank the enemy. I later learned that several vertebrae in my neck had been fractured when I took that hit.

"The last thing I remember about Flaherty was that he always carried an M72-LAW. It was an anti-tank rocket-propelled grenade launcher, and even though the Vietnamese didn't use tanks in the jungle, Flaherty loved to use it out in the field. I guess he got a rush out of blowing shit up. I'll always remember him as a solid leader and, for a small man, a great man to look up to."

––––––––––

Norm Campbell kept it short and sweet and wrote me the following in an email:

"When the Company started to form in late 1967, I was assigned to the 3rd Platoon and became the Radio Telephone Operator for Lt. Flaherty. I came to know him as a feisty leader who got along with the men very well but not so much with the other officers. He was well trained and seemed to know all about equipment, tactics and strategy. We were all heavily loaded with equipment and supplies, and he always carried his fair share despite his stature."

14

Special Forces

"Wherever the Special Forces operated, the CIA was close behind." — Anonymous Green Beret

1969 — Lopburi Thailand
46th Special Forces Company Base Camp Pawai

Twenty-three-year-old Captain Walt Yost is driven by jeep into the Thai Special Forces base camp. At six-foot-three and two hundred pounds, Yost easily springs out of the jeep and stretches his legs after the long one-hundred-kilometer ride from Bangkok.

He approaches a Thai guard and shows his identification.

"Where is Captain Flaherty?"

The guard, not speaking English, gives an acknowledging grunt and points towards a field a few hundred yards away. From that distance Yost sees what he believes is a small boy dressed in tiger-striped Special Forces camouflage talking to a group of Thai soldiers.

Yost patiently tries one more time with the guard. "Captain Flaherty, Amer-ri-can. Green Beret?"

The guard shakes his head up and down and points again to the small boy, who now appears to be conducting drilling maneuvers with the soldiers. Yost scratches his head and decides to head toward the boy. *Maybe he's Flaherty's son?* As Yost gets closer he realizes the boy is actually a tiny American man. He watches in fascination as the Thai soldiers obediently respond to the man's barking orders. The soldiers seem to not only have a lot of respect for the small man; they also appear very afraid of him.

"Captain Flaherty?"

Flaherty turns his attention to Yost.

"Captain Walt Yost, here to relieve you of your post."

Flaherty says something in Thai to the men and walks over to Yost. The two officers shake hands.

"Walt, it's a pleasure to meet you. Welcome to Shangri-La. If you follow me, I'll bring you up to speed."

The two men start walking towards Flaherty's makeshift office hooch.

"Intelligence should be sending out one of their handlers from Saigon to brief you on your objective," Flaherty says. "They'll be here within a week."

Yost asks, "They know what they're doing?"

Flaherty laughs. "You gotta be kidding me?!"

Walt laughs in turn. "Thought so."

They walk inside Flaherty's office. Flaherty sits down at a desk while Yost sits across from him. Flaherty pulls out a bottle of the Thai national beer, their version of Budweiser, which they call Singha (much stronger and far more bitter than any American beer — an acquired taste).

Flaherty also pulls out two paper cups. He fills them to the brim and hands one to Yost. The two men stand and raise their cups in toast, taking generous gulps as they sit back down.

Yost looks around the tent and sees a shelf filled with books: *The Art of War* by Sun Tzu, Homer's *The Odyssey*, *Meditations* by Marcus Aurelius, etc.

"We work any of our own intelligence gathering?" Yost asks.

"Unfortunately, that's a negative."

Walt leans toward Flaherty and quietly asks, "What's our cover story?"

"We're supposed to be training the Thai border patrol in reactionary force maneuvers. I have my guys work with them about ten hours a week."

Yost leans back in his chair and asks, "What are you going to do when you get back to the world?"

Flaherty grins. "Me, I'm a lifer. Once I get back to CONUS I'll transfer into training and try to pass on what I've picked up. The Army may be an ungrateful master and a real son of a bitch, but so far it has worked for me."

"You're a better man than me. I'm one-and-done after this."

Flaherty pours them both another generous portion of Singha. They both take long pulls from their now-soggy cups, savoring the bitterness.

Kosher Kingdom Bench

"What's the difference between what a Special Forces soldier does out in the field compared to a conventional soldier?"

"The mission is totally different. Special Forces troops are trained to operate always in enemy territory. You don't have a frontline. The actual mission of Special Forces is to infiltrate into an area, whether by air, sea, or land. Once in enemy country, you recruit indigenous personnel from that country to fight with you. You can recruit up to a regiment, which would be about fifteen hundred men.

"You first secure and then train them with weapons, which are usually resupplied from the outside. Once their weapons training is complete you train them in tactics.

"Then your responsibility is to conduct operations with them against the enemy government and government troops in that area. That's what Special Forces actually does. You're basically a guerrilla.

"That's why you don't think like a conventional officer, you think like a guerrilla. You're not going to have any friendlies around you, you're going to be surrounded all the time. You've got to be able to survive out there on your own. You have to change everything the Army trained into you during boot camp ... and beyond. No more following the rules and no more going by the book.

"Back in those days we communicated by CW, continuous wave radio, which is dots and dashes, Morse code. Everything was crypto-code; it's a one-time codebook. You use a page, throw it away and destroy it. It was very advanced for its time. Now it's like bricks and rocks, but back then that was cutting edge stuff.

"Special Forces, specifically an A-Team, doesn't have a whole lot of infrastructure. It's just them out there and what they call the SFOB, which is Special Forces Operating Base; that's what and who they communicate with. The twelve-man A-Team recruits up to fifteen hundred indigenous personnel, so you're basically operating a full regiment with just twelve men.

"But each man of that A-Team is a leader in his own right. If you went down someone on the team would know how to take control, do your job. That's why an A-Team can operate that big of a unit, because all the military disciplines needed are within each man. You could actually split the team in two and have the six specialties on each side.

"There was also a higher level of physical ability expected of you. You have to be able to do everything, be able to go through any type of obstacle, and if it wasn't something you could physically overcome — you had to mentally overcome it.

"How do I get around this obstacle, how do I figure out a way to get this unit from here to there even though they've got something in the way? And so, all of the men in your team would also have to be physically capable of doing that. An A-team is just like a chain, the weakest link makes every other link vulnerable. You can only move and operate at the efficiency of your weakest person. So, when you're in a Special Forces team it's a brotherhood. It's not so much about the officer-to-enlisted-man relationship anymore.

"It's like we're on the same team: I may be a captain, you may be an E5 sergeant, but we all know our jobs and we're going to do these jobs. Even better, when we're not operating you're going to teach me and all these other guys your job — and vice versa, we're going to teach you our jobs.

"That's what we did in the team rooms: we cross-trained constantly. So, I would know how to operate the radios, I would know how to do medical procedures, I understood all aspects of the heavy and light weapons, I was capable of setting demolitions — and they all understood all that too. On top of everything else add knowing all the operations and intelligence stuff."

15

The RIF

"Betrayal is the only truth that sticks."
—Arthur Miller

1971 — North Carolina
U.S. Army Fort Bragg

Flaherty is sitting in his well-furnished office, finishing off his daily stack of reports. On the top of his incoming mail pile he spots a U.S. Army letter. *This better be my transfer into training*, he thinks as he opens it. He scans the interior of the letter and reads:

> Captain Richard James Flaherty, your contract with the United States Army will not be renewed. You have 30 days to return all Army issued equipment and complete your decommissioning paperwork. Please arrange a time with your supply sergeant to get checked off on all items pursuant to your dismissal.

Flaherty's hand starts to shake as he stands up and bolts out of the office.

With his jaw tightly clenched he marches into Major Mac Freeman's office. Following on his heels with a worried expression is the major's secretary, Ms. Redman.

As Flaherty barges into Freeman's office, Ms. Redman announces, "Sir, I tried to stop him but ... "

Freeman puts down some paperwork and calmly tells her, "That's okay, Ms. Redman."

The secretary leaves the office while Flaherty stands rigidly, the letter clutched in his hand.

Seeing Flaherty's face, Freeman reaches into his desk drawer while telling Flaherty, "Captain, would you please have a seat."

"With all due respect, no thank you. What's the meaning of this?" Flaherty waves the letter like a flag, paper crackling.

Freeman takes his Alka-Seltzer bottle out of the drawer and plops two tablets in a half-filled glass of water. He doesn't wait for the Alka-Seltzer to melt, downing the glass with a couple of crunches. He holds out his hand and Flaherty passes him the letter.

After a quick scan he exhales a deep breath and says, "Come on, Richard. You knew damn well this downsizing was coming."

"What happened to at least offering me the chance to drop down a rank? I'll go all the way back to a second lieutenant if I have to. Major, this is all I have."

Third officer this week waving his letter in my face — and this is just the beginning. Freeman grimaces as an acid belch churns in his stomach.

"Son, I've read your jacket many times. I know what you've done for our country, but these new politicians promised the taxpayers a tighter budget, a leaner military, a reduction in force, you know the RIF. You know the first to get squeezed out are always going to be the captains and lieutenants. Welcome to the new fucking Army!"

What a kick in the stomach. I looked at Richard and felt his pain. I shook my head and said, "Damn. That must have felt like the ultimate betrayal."

"I would have done great things for this country. And it was more than just the end of my career. Outside the military I was no longer Captain Flaherty. I was just that little guy, that dwarf again." He bitterly spit out those last two words. "All that I earned was taken away, forgotten."

"Sorry, brother." What else could I say? It was like notifying a family member of the death of a loved one and adding, "Sorry for your loss." It somehow never seems to be enough.

"When I got home they threw me a huge homecoming party at Pellicci's restaurant in downtown Stamford. Nice party, all my family and friends showed up. Everyone was drinking and happy. I got a chance to play the role of a war hero for the first time, and wore my full class A uniform with all my salad dressing — that's medals to you civilians," he added wryly.

"It was the only time I saw my dad real proud. He kept walking me around the room saying, 'That's my boy. That's my son.'

"I even got along with my brother Walter. I never saw him drink 'til that night. For once he dropped his tight-ass thinking-he's-my-father attitude and just drank and laughed with me like a big brother's supposed to. Anyway, at the party there was an old friend of mine from P-Town."

Richard saw the confused look on my face and asked, "You know P-Town?"

"Never heard of it," I admitted.

"Oh, P-Town is short for Provincetown Massachusetts. It's a small seaside town that was once known for being the landing site of the Mayflower. Now it's known as the landing site for hippies, artists, and all-around nut jobs.

An entertaining little town; before I went into the Army me and my friends would head up there on the weekends.

"Anyway, my friend Abe drove down from P-Town for the party. He told me he was making a killing up there managing several clothing shops. He asked me if I wanted a manager's job, or maybe we could open another clothing store together. That's when it hit me to try my hand in the garment industry. I already had some connections with garment factories in New York City, so I figured I'd give the manufacturing world a go."

16

A Girl Named Jane

1973 — New York City

Twenty-eight-year-old Richard Flaherty (wearing a three-piece suit) is sitting in his office, looking through a mountain of bills and wondering how he's going to possibly make next week's payroll.

Boy, did I bite off more than I could chew. As he ponders his future he reaches for his reserve bottle of vodka, stored safely in the lower desk drawer.

A gentle knock on his half-open office door draws his attention.

"Come in."

A petite woman, dressed in a business suit and holding a briefcase, hesitantly takes a step inside.

"Mr. Flaherty, may I come in?"

Richard is stunned, and then he remembers his manners.

"Sure."

The woman steps to his desk with an outstretched hand and says, "I'm Jane Cohen, with Omega Fashions."

Richard comes quickly around his desk, takes her hand and guides her to a cushioned chair.

"Jane, have a seat. And please — call me Richard."

Kosher Kingdom Bench

I witnessed emotions I rarely saw from Richard as he told me the story of meeting Jane: happiness and excitement.

"Only a few weeks in and this saleswoman walks into the office. Dave, she was beautiful. Long blonde hair, a face like an angel, and just a drop taller than me. She was selling fabrics. Right off the bat we hit it off like a house on fire.

"My boy, some romances take time to develop. They keep you taxiing on the runway, waiting for clearance. Then, there are romances that take off like a jet with after-burners blazing — have your head so quickly in the clouds you're reaching for your oxygen mask. That's the way it was with Jane and me that day in the office: we just clicked.

"We did it all. Horse and buggy rides through Central Park, picnics in upstate New York, and we even saw Led Zeppelin give one of their first concerts at the Garden. Everything was going along great until later that year, when she invited me to her parents' house for Thanksgiving."

1974 — Hoboken, New Jersey
The Cohen House

Inside the dining room of the Cohen house, the family is all seated at a large table.

The room smells of fresh-baked gingerbread and turkey stuffing. Flaherty, the guest of honor, sits just to the side of Jane's father. Jane sits next to him, nervously holding his hand. Across the table is Jane's mother and her younger teenage brother, Todd. Although the rest of the family is conservatively dressed, Todd is wearing a tie-dye-colored T-shirt and faded jeans. He sports a long and unkempt hairstyle with a beard to match.

110

Flaherty takes a break from the simple "Where are you from?" conversation to glance around the room. A hand-painted wooden plaque stationed above the entrance to the kitchen reads *The Cohens* — lovely and simple. Almost all the wall space is covered in framed pictures of the entire family throughout the years. *Man, they're a close-knit group.*

Halfway through the meal a mild discussion of world politics starts up between Jane's father and her brother Todd. Jane's father says, "Nixon was supposed to get us out of there by now, and who knows? Maybe he's right about the bombing campaigns."

Jane's mother shifts uncomfortably and puts her hand gently on her husband's arm. "I thought we had a rule about politics at the dinner table, especially when we have guests."

Todd pays no attention to his mother and smiles at his dad. "I'll tell you what I think, I think Jane Fonda is a fucking hero. She's the only one standing up to those warmongers, and ... "

"Um, mom," Jane cuts her brother off. "Did I tell you Richard and me are working on some new designs for the factory? We ... "

Todd ignores his sister. Pointing to Flaherty with an accusatory fork he asks, "Hey Ricky, what do you think about the war?"

Flaherty calmly smiles and puts down his glass, wiping his mouth with his napkin before replying. "Well, that's a complicated question requiring a complex answer. Not everything that you're hearing on the news is the truth. You really have to be in-country to understand all the issues at play."

"Oh? How would you know what's really going on over there? You get your news just like the rest of us," Todd replies snidely.

Jane nervously clears her throat and says, "Mom, Dad.... I probably should have told you earlier, but I never got a chance because everyone is so caught up in emotions, and ... "

"Honey, what is it?" her mother asks.

"Well, Richard just got back from Vietnam. He was an officer, a captain in the Army, and he told me that the real truth is ... "

Todd stands up abruptly, knocking his chair backward, and levels a finger straight in Flaherty's face. Horrified he shrieks, "A FUCKING BABY KILLER? YOU BROUGHT A FUCKING BABY KILLER INTO OUR HOUSE! Fuck this! I'm not sitting here with this war criminal!"

Todd storms out of the room. The silence in his wake is deafening.

Jane's mother finally breaks the uncomfortable quiet, leaning toward Richard. "Richard, I'm very sorry about Todd; he's just so passionate about the war." She turns to her husband and pleads, "Jack, please tell Richard he's our welcome guest. Todd was just overwhelmed by all the tragedy we've been seeing on the news."

Jane's father stares at Richard with disdain, then starts eating his food in silence. Jane looks to her father, pleading with her eyes for him to offer a little civility to her boyfriend.

"Thank you for the dinner, but it's probably best I get going," Flaherty says. He somberly gets up from the table and exits the house.

Flaherty walks down the driveway towards his car. Jane, crying, runs after him.

"Richard!"

He stops and turns around; she runs headlong into his arms. "I'm so sorry," she whispers.

"It's alright. It's not your fault. Feels like the whole damn country has turned on us. Look, I had some news I was going to share with you after dinner, but now seems like a good time. My old friend Abe Saada told me he would bail me out and take over the factory."

"But I thought you loved that place!"

"Are you kidding me? I hate it — the only good thing that's come out of it has been you. If I sell, we're finally free to move to Miami."

"Do you mean it? Can we really go?"

"There's nothing holding us back: it's just you and me now. It's time we start living on our own terms."

Jane joyfully hugs and kisses Flaherty.

Kosher Kingdom Bench

"We were really happy at first when we got to Miami, but then we hit a rough patch. Jane couldn't get the job she wanted. And I started having trouble dealing with things — normal things that shouldn't have ticked me off, but they did. There were times I would leave the apartment in the morning and walk until the sun was setting, not even knowing where I was going."

"Sounds like the early signs of PTSD."

"Back then we didn't have terms like PTSD. They called it shellshock or battle fatigue, and it was something that only happened to the weak-minded — or so we thought. It freaked Jane out 'cause she didn't know what to do to help. I also knew she missed living close to her family and having that support. We started doubting if we made the right decision.

"And then things finally turned around. She got a teaching job that she loved, and I started working for an old Vietnam vet friend designing and selling rifles. I thought things were finally looking up for us."

March 4, 1975
Miami Beach Apartment

Inside Flaherty and Cohen's one-bedroom apartment, Richard is finishing off some new weapon designs that he plans to pitch to his boss at Gwinn Industries. A knock on the door brings Flaherty off the living room couch. He opens the front apartment door to see a nervous-looking Miami Beach police officer standing just outside.

"Can I help you?"

"Are you Richard Flaherty?"

"Yes sir, I am. Is there a problem?"

The officer uncomfortably shifts on his feet and solemnly says, "I'm sorry sir, I have some bad news about your fiancée. I need you to come with me."

North Miami Beach N.E.
163rd Street and 34th Avenue

A crime scene with yellow tape surrounds a car that is folded up like a crushed can around a utility pole. An ambulance is parked next to the scene. Flaherty and the officer are standing by its rear door. A paramedic opens the door and removes a sheet covering a body lying on a gurney.

"Is that her?"

"Yea."

Stunned, Flaherty walks away from the scene and sits on a wooden bench facing the Intracoastal Waterway Bridge. After a few minutes the Miami Beach officer approaches Flaherty and states, "Sir, there was a witness to the crash. An off-duty police officer who was driving home. He was a few hundred yards behind her when it happened. He said she passed him at a high rate of speed about a mile earlier, and for no reason that he could tell just lost control of her car and swerved straight into that pole."

"Nothing in the road? No debris? An animal crossing?" Flaherty asks.

"No sir, or at least he didn't see anything that could have been a contributing factor. By the time he pulled over and got to her she already had no pulse. I guess we will have to wait for the toxicology reports to come back. Maybe we can check the vehicle for defects. I don't know. I'm not the lead detective on this."

1975 — Miami Beach Apartment

Flaherty's apartment is in complete disarray. In the living room the blinds are broken and hanging sideways, allowing the morning sun to flood in on empty liquor bottles and moldy pizza boxes littering the floor. The once-immaculate kitchen overflows with piles of unwashed dishes and accumulated bags of garbage. The hallway leading to Flaherty's bedroom is a cacophony of discarded clothes and other trash. In the dimly lit bedroom Flaherty sits on the edge of the bed, an anguished look etched on his face.

It's been a week since he identified Jane's broken and bloody body. Her funeral is scheduled for the following day in New Jersey, but he knows he can't go. *If they hated me then what do they think of me now? It was my job to protect her, and now she's gone. Another life I was responsible for, gone.*

Flaherty opens another bottle of vodka and it's not even noon. *How long can a man live on vodka? I guess I'll see.* The alcohol gives him no comfort, but he continues to self-medicate, hoping it will eventually grant him relief — or at least a break from the nightmare that consumes him. Even sleep evades him, except for fractured blackouts of time. He is wakened from his latest blackout by an irritating sound coming from the bathroom.

He stumbles off his bed and heads towards the bathroom. As he navigates the dark hallway the persistent drip-drop, drip-drop, drip-drop of a leaking faucet mixes with the gurgling sound of water going down the drainpipe. As he enters the darkened bathroom the sounds increase in volume, so loud he can feel them pulsating through his body.

In frustration he smashes his hand upward onto the wall light switch. The light reveals the truth: the sink and shower are bone dry. The sound was always coming from deep inside him. Flaherty crumples to the cold tile floor and clutches his friend Sergeant Bill Meeks, incoming bullets chewing up the jungle floor around them.

Where the hell is that medic? Meeks is losing too much blood, I can't stop the bleeding. Meeks' blood all too loudly drip-drops onto Flaherty's canteen as he applies pressure to the sucking chest wound. The drip-drop is replaced by a loud gurgling sound, Meeks' blood pouring down his throat. The man rattles, twitches, and draws his last breath. *Another life, gone.*

The next morning, before sunrise, Flaherty heaves the third garbage bag full of empty bottles and refuse out of his apartment. The sun is just starting to rise, and the air is crisp and clear. Flaherty heads toward the beach for a morning run.

As he jogs on the sand, breathing the fresh salty air, he looks toward the horizon and watches the sun emerging from the cobalt-blue ocean. Jogging always helps Richard regain his perspective, and he needs a new perspective now more than ever.

It's all about perception. Miami has its different moods, just like Vietnam. In the mornings everything is so vivid and clear. By mid-day the harsh sun washes out the colors, making the world dull and hazy. But in late afternoon the sun provides a warm glow, renewing a man's spirit. The night ... well, the night conceals all ... it's time to embrace the night!

17

The Seventies

*"Stuff that's hidden and murky is scary
because you don't know what it knows."
—Jerry Garcia*

Kosher Kingdom Bench

Richard wouldn't go into any further details of what he did
after Jane's death. All he would tell me was that he was
in a dark place that reminded him of being underwater
in that Vietnamese rice paddy. Unfortunately for Richard,
this time there were no teammates around to reach into
the dark waters and help pull him up. The only thing he
would say about that time was once you hit rock bottom
there's nowhere else to go but up.

Some of my leads into Richard's activities came in the
form of a two-page typed document buried in his rusted
file cabinet. The document was titled Resume by Richard
Flaherty. In his resume Richard wrote that he worked for
Gwinn Industries from 1973-1990.

My research revealed that Gwinn Industries was a
weapons company founded by Mac Gwinn Jr. The company was eventually sold and turned into the now-familiar Bushmaster rifles. I remembered Richard briefly mentioning the company Bushmaster Rifles, saying that he
worked for them throughout his life.

I contacted Gwinn Industries and spoke to Mac Gwinn III, the son of Mac Gwinn Jr. He told me that his father was an Army Special Forces soldier who served in Vietnam. He added that he remembered as a boy meeting Captain Flaherty.

"He was a tiny guy, but he was real muscular. I know that he worked with my dad."

He was unsure exactly what Flaherty did for the company besides some sales — and possibly helping develop and test new weapons.

Another document I found interesting was a letter Flaherty wrote to Ross Perot regarding some new type of weapon he had developed. He was looking for Perot to come in as an investor.

"Rich can you at least give me some idea of what you were up to?"

"Look Dave, there were times in my life that just seem like a blur; this is one of them. What I can say for sure is I needed to find some work, so when I saw this help wanted ad in the local paper for a used car salesman I decided to give it a shot. I called them up and scheduled an interview for the very next day. So, I go in for my interview and they had this real tall front counter. I could barely see over it.

"The front desk girl asked me my name, and because I could barely see her I kind of loudly yelled it out. The next thing I hear is a man's voice from the back shouting out, "Flaherty?"

"Well, wouldn't you know it but my old friend from Thailand Green Beret Captain Walt Yost comes running up to the counter. Walt had been working there for about a year, and that was that: I was hired on the spot."

I was able to locate Walt Yost through a website for the 46th Special Forces Company retirees. I emailed the webmaster and explained who I was and what I was doing. He gave me Walt's phone number.

I called Walt, and he was very interested in meeting up with me. According to him Richard had been a good friend; Walt was sorry that they lost touch over the years. Walt told me he was about to head up into the mountains of North Carolina — did I want to meet him up there to talk about Flaherty? I had never been to those mountains, so I figured: why not?

I met with Walt on the shores of beautiful Lake Nantahala for our interview. It was an incredible day with clear blue skies, around seventy degrees. Walt already had a cold beer waiting for me.

My interview with Walt was interesting but brief. Besides describing a few days together in Thailand he didn't have many Richard war stories. When it came to explaining his mission for the 46th Special Forces Company Walt stayed exactly on script, reciting almost word for word what Richard told me of their mission parameters. I then asked him about working at the car dealership with Richard, and got a similarly apathetic response. "You know, selling cars is just mundane type of work. Nothing too exciting."

He finally opened up a little bit when we talked about what two mid-twenty-year-old guys did for fun back then in Miami Beach. According to Walt there were some wild nights out on the town, which included heavy drinking. He talked about a bar fight that was started by a drunk guy picking on Richard. Walt said he sat back and watched Richard quickly handle his business, making the much larger man enjoy the taste of the barroom floor. He also remembered that night because Richard got so drunk he had to carry him to the taxi.

"How did you carry him?"

"You know the way you scoop up a small child and cradle him? I mean he just wasn't that heavy!"

I kept in touch with Walt after the interview, and I would call him every now and then to say hi and shoot the shit. One day I couldn't take it anymore and flatly asked him what he and Richard's real mission was in Thailand. Walt, like any good intelligence officer, refused to budge on their team being involved in any activity besides training the Thai Royal border guards.

I then questioned him on the coincidence of two of America's most highly trained and experienced Green Beret Captains accidentally ending up working for the same used car dealership in Miami. Walt, without missing a beat, calmly answered, "That's the way it happened. It was one of those crazy coincidences."

Over the next few phone calls I kept pressing and chipping away at Walt, and he kept answering in his most innocent voice that things really happened that way. One time we were on the phone laughing about something, and I squeezed him again for the truth. Finally he cracked and said, "Yeah well, maybe the car dealership job wasn't so much of a coincidence." He would go into no further detail of who placed them there, but he revealed that the dealership was basically a jumping off point for clandestine missions.

He went on to tell me that in the late 1970s he and Richard were approached to start gathering intelligence and give an assessment of the type of mercenary team needed to invade Grenada and overthrow their government. Walt refused to tell me who approached them or go into any further detail, except to say that the project was eventually scrapped. It should be noted that on October 25, 1983, the United States military invaded Grenada under the codename 'Operation Fury.'

I think I only spoke to Walt one more time after that by phone. In that call, he talked in very abstract terms about

Richard doing some type of advisory work in Rhodesia and South Africa. Sadly, shortly after that conversation Captain Walt L. Yost had a heart attack and passed away on March 8, 2017. Captain Yost was a recipient of the Bronze Star and a Purple Heart for his actions in Vietnam. Rest In Peace my friend.

18

Entrance to the Rabbit Hole

During our research my father and I were given the name of a man who was supposedly close to Richard throughout the seventies. My dad was able to track him down, and after many emails and phone conversations managed to gain the man's confidence. They eventually met in Malibu, California for an interview.

For his protection, we are using a pseudonym — Frank Sosa.

Frank would only refer to Richard as Rick, claiming that Richard would only answer to the name Rick back in those days.

They met at a local coffee shop and my father taped the interview. The following excerpt is narrated by my father Neil Yuzuk.

Coffee Shop Patio — Malibu

It was a cool but sunny day as we sat on the coffee shop patio. Frank did not want to talk inside, although he agreed to allow me to tape the conversation. Frank was a slightly-built man who (despite his claims of life-threatening illness) looked healthy as he constantly pulled at his e-cigarette. He wore sunglasses and a cap like a spy film informer should, and he initially moved somewhat furtively in his chair. As time passed (alongside several cups of coffee and a lot of vaping) he became more relaxed. The interview took three-plus hours; I've deleted those portions that I used to establish commonality with him.

A strong handshake and a brief "Hello nice to meet you" out of the way, Sosa quickly got down to business. "First and foremost, okay, say what you want about Rick, about him having some twisted traits and being involved in some shady things, but he was a patriot before anything else. The other word I would use to describe a man like him is a renaissance man."

"How did you meet Rick?"

"In the early seventies I lived down in South Florida and worked for a man named Norm Lalo. Okay? And Norm was a bastard of the highest repute. Norm owned several gun shops in Miami and Broward County, and I started working for him when I lost my electrician's job.

"Norm's shops attracted the most colorful of characters — basically a pirate's cove, except these men were far more dangerous than William Kid or Calico Jack. You have to remember that this was Miami in the seventies: Cocaine Cowboys, Scarface, money laundering, drugs, guns, and then even more drugs, you name it.

"Me and Rick first started socializing when he would come into the gun store. Later on he invited me to his house for these dinners and parties. The coral rock house in South Miami was a small guest house that Rick was living in at the time.

"There was this other guy who came into the shop that hung around Rick and came to his house parties. He was also a mercenary in Rhodesia, among other places in Africa. A Vietnam vet, who eventually ended up in Leavenworth for killing people. Okay? His claim to fame was he would use an aluminum baseball bat with a plastic cover over it, which is how they used to come when you would buy them back then. He'd take that bat and beat you to death with it, drop the cover someplace it would never be found and then take the bat and throw it away somewhere else.

"They finally caught up with him. He was doing it for — he would say corporate assassination, but it was probably hit jobs for various agencies from around the world where he had been. I'm just sitting there in the shop keeping my mouth shut listening to these guys talk.

"We also had some undercover narcotics cops that used to hang around the store. You always knew the undercover guys because they made themselves look scummier than the actual drug dealers. Then there were a couple of outlaw bikers that would stop by, a guy named Ed who had a machine shop down in Davie. He eventually ended up doing federal time because he was making silencers. He used to buy Beretta 25s from the shop for who knows what, and then...." Sosa paused. "Sorry for rattling on. I do that often."

"No problem."

"So, give me some idea of what you want to know."

"Let's start off with, do you think Richard considered you a close friend?"

"No, no man. Rick never befriended anybody in his life. He just never fully trusted people. Look, Rick was a little paranoid. Now just 'cause you're paranoid, that doesn't always mean you don't have enemies! Some of those enemies could be enemies that you self-made. Okay? You know Neil, you said something to me on the phone about perception. That everybody who knows Rick sees him in a different way."

"Yeah, I was saying how everyone sees him in a different light."

"No man, that's wrong. You are one hundred and eighty degrees out of phase, bro."

"You think?"

"Yeah, totally. You see, people aren't seeing Rick differently. It's Rick doing that on purpose. See, Rick presents a different version of Rick to everyone he meets. You get how that's different?"

"I believe I was saying it more ... "

"It's a different thing man, completely different. Let me tell you a quick story. Okay? My best friend Ken Myers was a cop and detective in Miami, and Rick goes to him one day and says, "Hey, you keep saying you're a good friend of Frank's."

"Ken goes, "Yeah, he's my best friend."

"So, Rick continues with, "Well, let's just see how much of a friend you are, and how much of a cop you are. What if I told you I have a gun with Frank's fingerprints all over it, and that gun was used to kill a man, and the ballistics will prove it, and I can give you that gun?"

"That's what Rick did to everybody. He tested them.

"You know what Rick was like? You remember that Mad magazine comic strip Spy vs. Spy? That's Rick. When he'd leave his house, he'd do little telltales on the front door to check later to see if anyone went inside. When he parked his car, same thing. It was like he was on a mission — 24 hours a day, 7 days a week, 365. Nobody can live like that, man. Nobody."

Sosa took a long pause and looked away. He nodded his head as if in conversation with himself, then turned to me.

"Here's the sad part — this is what used to tear me up, man. We'd be sitting out on the boat. He liked being out on my boat off the inner coastal, out in the bay 'cause he felt comparatively safe out there. He'd make me kill all the lights except a couple running lights which you have to have, otherwise you might get run over. This way he could see anybody coming up on him from any direction, and that would give him comfort. We were usually eleven miles offshore, off of Turkey Point right near the power plant.

"Anyway, that's when he would break down and start

telling me things. Again, I keep using the same word: frustrated. Here's a guy that could talk to you on any subject or issue you could imagine. He could talk to you about John Locke, Rousseau, Greek philosophy, various religions, all types of art, just amazing, and a great cook by the way. I don't think most people know that, but Rick could really cook up some nice food.

"In the few times I was with him when women were involved (if you know what I mean), the women seemed really happy afterward. So, he was probably pretty good in that area too. That's what made it so sad, because if they'd given him some opportunity — they got a gem here. I think his life would have gone ... "

"A different direction, maybe?"

"Let's not go that far, but at least veered, maybe not done so much of that crazy shit. I am convinced that it was those nights on my boat that.... I don't want to say he warmed up to me 'cause I don't know that Rick warmed up to anybody. There's always a big piece of Rick that is just never going to trust you."

After a few moments of silence I asked, "We heard that Richard might have been doing some mercenary work in Africa, but his passport doesn't show he ever went there."

"Let me put it this way," Sosa laughed. "When we went to Africa I didn't even have a passport. Rick knew ways to get in and out of the country that were undetectable even to the State Department."

"Would you talk to me about Rhodesia?"

"Well, let me first start by telling you Rick would've objected to the word 'mercenary.' Now let's get this straight — did Rick throughout his life get paid to fight in numerous conflicts around the world? Damn right he did. But he always picked his battles, and he never allied himself with any government or system that he felt was against the interests of the United States ... and he fucking hated communism. If the Russians or Chinese were sending advisers or soldiers to some corner of the world Rick would want to be there to oppose them.

"He told me the oldest profession in the world — and the true prostitutes — were mercenaries. Men for hire who kill indiscriminately for money. Forget the Soldier of Fortune romantic lore about young men going off to the Foreign Legion and seeking adventure — that's not reality. The reality is most mercenaries are hired to do the dirty work, the butchery that a professional soldier won't do. Now I'm not talking about private military firms or contractors, executive protection, or even the knuckle draggers."

"Knuckle draggers?"

"Knuckle dragger was a term we used to refer to any independent contractors who were paid by the CIA to do field work. The guys I'm talkin' about were the ones with no loyalty or morality, who would switch sides in the middle of a conflict just as long as they were offered more money.

"When Rick and me went to South Africa we were recruited by a private military firm. They offered us a cash signing bonus when we landed in London, but Rick wouldn't let me sign any contracts until he went in-country and checked it. We landed in Cape Town, South Africa — man, what a beautiful country. It's truly breathtaking. We were supposed to start a thirty-day adaptation training once we landed, but Rick had other ideas.

"That week Rick met up with some other Green Berets that he served with in Nam and they gave him the real skinny. Okay? He found out that a lot of the private military firms over there were scamming guys out of money and not paying what they were promising. And the equipment and uniforms they provided were used and in poor condition. Also, most of the company executives — who didn't even have a fraction of the experience of Rick or these other Green Berets — were the field officers and supervisors, and they wouldn't relinquish their power to anyone else. So, we turned down all the offers and flew to Rhodesia on our own dime.

"Now Rhodesia was just as beautiful as Cape Town, but the landscape was more virile, untamed. Rick had contacts in the Rhodesian SAS and he ended up getting us hired directly by the Rhodesian Army as advisers. He even negotiated our contracts. We were initially hired on a six-month contract which included all kinds of compensations, like if we got wounded, hospitalized, or killed. We also received danger pay anytime we operated outside of our base camp. As soon as we signed our paperwork they handed us bottles of Camoquin. Camoquin was a drug that supposedly stopped the biggest killer in Africa, which was malaria.

"The first thing Rick did when we got into base camp was to talk to some of the top trackers from the Selous Scouts. The Selous were the Special Forces of the Rhodesian Army that basically handled all the clandestine missions. See, Rick was already a master tracker; however, he was now playing on a new ballfield. He wanted to learn everything he could about the terrain, weather, foliage, and animals in the Rhodesian Bush. The next thing he wanted to see was what type of footwear the guerrillas preferred. Many expert trackers will keep books that include hundreds of detailed drawings or pictures of shoe soles.

"Remember I was never a soldier, but I was a weapons expert and gunsmith. I could also competitively shoot with any man back then, and still to this day. So, I always worked with Rick on teaching him about the parts and nomenclature of hundreds of weapons, and he would teach me about soldiering. When we'd head out to the Everglades Rick would teach me dozens of things to look for when you're tracking a man.

"He'd always start with the terrain and atmosphere. He'd just stand there and tell me to observe all the sights and smells. Now if the sounds change or stop it's an indicator that someone is or was recently in the area. Okay? Then he'd start looking at the smaller things, like spider webs to see if they were recently broken. He'd look for bent or broken blades of grass. If it was bent then the direction of the bend would indicate which way your subject was going, and if the blades were broken they would only stay green for about a day after they were damaged. He'd have me look for rocks that had been recently overturned. You could easily spot them because the rocks will have the darker side facing up. The darker and damper the rock, the closer you are to the enemy.

"He also taught me techniques on how to defeat someone tracking you. To make sure you walk on the hardest ground you can find and utilize rivers and creeks whenever possible. To try to take advantage of rain and inclement weather to hide your track. To try to wear the same footwear as your enemy to confuse him. When traveling on soft terrain remember to always brush out your tracks with palm fronds or ferns. Last and most importantly, always booby trap your trail to slow and discourage pursuit.

"Now getting back to Africa ... let me first give you a little history lesson about what was really going on over in Rhodesia. You can't believe everything that came out of Walter Cronkite's mouth back then, and the newspapers were all running stories about Rhodesia being this rogue evil nation of white minority rule that was oppressing the blacks. But that wasn't true, and I'll tell you why. Okay? Because Rhodesia — which was isolated by the world with import embargoes and surrounded on all sides by enemies — should have fallen in a couple of years, not lasted over a decade.

"The problem the Nationalist groups calling for a majority black rule had was they didn't have the support from the people. The first rule in any guerrilla war is the majority population must support the conflict. Africa has always been an unstable continent, ruled for God knows how long through a tribal hierarchy. There were just too many blood feuds going back hundreds of years that wouldn't be ended by someone just handing them democracy and saying 'good luck boys, have at it.'

"The reason the majority of black Rhodesian citizens backed the minority white government and voluntarily fought in the military was because they were safer — and their standard of living was higher — than any of their neighbors. It wasn't a perfect situation, I get that, and of course they wanted more rights, but they were more interested in their children's future than immediately seizing power. The plan was for a much slower transition to a majority lead black government, but the world had different ideas.

"So here comes old Russia and China again to back the Nationalist's movement and provide them with weapons, funding, and training. Okay? The tactics were the same guerrilla tactics taught to the Viet Cong that Rick and those other Vietnam vets over there had already seen. This was never a conventional war with big battles — there never was a front line. Neither side had much in the way of air power, and the helicopters and planes were mostly used for supply and reconnaissance. Rick and I mainly stayed in the base camp to provide training to the new recruits, but I do know on a few occasions Rick just couldn't help himself and snuck out with the troops on search and destroy missions.

"Every day you'd hear about a bus hitting a land mine or a bomb planted in a petrol station being detonated. Like the VC they had very crude base camps, they built no military infrastructure, and they never attempted to occupy territory.

"The year I was in country there were increasing incidents of Rhodesian farmers living on the outskirts being horrifically slaughtered by the guerrillas. The commander of the guerrilla army, Robert Mugabe, would quickly denounce the attacks and state in the press that those were just rogue terrorists, but we knew different.

"There was a family that me and Rick met that owned a farm just outside Bulawayo. Bulawayo was the second largest city, and sat closer to the border of Botswana. We had swept the area once and met all the local landowners to talk about security. This one family had invited our entire squad in for lunch. Very lovely people, and the farm employed maybe twenty of the locals.

"The guerrillas waited that night until we left the area and then slaughtered every man, woman, and child on that farm. They shot the owners of the farm and their family execution-style in the back of the head, but the black employees were brutalized in unspeakable ways before they were killed. They chopped off several of the children's heads and stuck them on the ends of assegai spear tips. The guerrillas were sending a message to the local population that if they worked or cooperated with the whites they would be seen as traitors and treated as such.

"The next day after we heard the news our entire squad wanted to go into the bush after them, but the commanders ordered our unit to stand in place as security for the other landowners in the area. Rick wanted to go in alone, but two other Crippled Eagles who also wanted revenge defied orders to go with him."

"Crippled Eagles?"

"That was the informal name of the American Vietnam vets that served in the Rhodesian Army. These were the men that also had their careers shortened by the reduction in force, and the name Crippled Eagle was a dig at the country for betraying them.

"Rick and the Eagles were gone for eleven days. Most people were writing them off, but they finally made it back. They carried Rick on a makeshift litter, and he was hospitalized for a couple of weeks."

"Was he shot?"

"No, he was running a high fever and had infections all over his body. He himself never told me the story of what happened. I got it from the other two American soldiers. They told me by the third day of tracking they found the guerrilla's base camp. It was a pretty well-defended camp, with fencing and guard posts, containing approximately seventeen of the guerrillas.

"The two soldiers wanted to immediately hit the camp, but Rick made them wait two more days so he could observe it and come up with a plan. On the sixth night, Rick inserted himself into the camp by crawling through a small creek and up a dug-out dirt trench that ran under the fence — which the guerrillas used as their latrine. There was a sentry nearby, but he was watching the outer perimeter and not guarding the water supply, which was stored in five-gallon steel gasoline cans in the interior. Rick poisoned their water with some type of concoction he made from pieces of rotting dead animals mixed with animal droppings and gun cleaning solution.

"Rick and the men then waited and watched the camp for another full day. By the next day half of those guerrillas were suffering from some form of severe dysentery. That night Rick crawled back in through the same shit trench and laid in that foul muck. He carried flares, grenades, and a .22 Caliber High Standard semiautomatic pistol fitted with a silencer. It's a real quiet weapon, but it has almost no stopping power, so you need to get really close to the enemy for a clean headshot to put him down. The other two Americans, carrying Belgian FN FAL rifles, placed themselves in firing positions high up in sturdy trees that overlooked the entire camp.

"Rick killed two guerrillas that came by the ditch to do their business. The first guy was a clean headshot; Rick quickly pulled the man's body into the muck and waited for the next guerrilla. The next man wasn't a clean kill. A nearby guerrilla. heard the dying man making noises and sounded the alarm. The most important part of Rick's plan was to set flares off inside the interior to create a well-lit shooting gallery. Rick quickly threw out several flares in all directions, then tossed a grenade to cover his back as he hauled ass out of there. By the time he got back to his own firing position — which was maybe only a minute or so later — the fight was already over. Three or four of the guerrillas escaped into the bush, but the rest were ground up by those FAL rifles. With the camp illuminated and half the guerrillas so sick they were barely able to run, it was like shooting fish in a barrel.

"They did interrogate one wounded man and got some good intel. The day after the attack, as they were making their way back to our camp, Rick started running a high fever and collapsed. It's obviously never a good idea to lie in a pool of shit when your body has hundreds of small cuts and bug bites from traveling through the bush.

"When they did finally get back all three were in pretty rough shape from lack of food and water, with Rick of course being the worst. The Rhodesian commanders understood and were very thankful for what Rick and those Crippled Eagles did, but they also had to make an example out of them for disobeying orders. The two Eagles who were sworn into the Rhodesian Army were fined a month's pay and sent back to the capital in Salisbury. Rick was already on thin ice with the Rhodesian high command because a few weeks earlier he'd bumped heads with a Major General. But Rick was too valuable to not be utilized, so instead of sending him home as soon as he got out of the hospital they sent him off to Angola to continue as an adviser. I stayed on as an armorer for the Rhodesian Light Infantry for an additional six months."

"Did Rick ever talk to you about Angola?"

"The next time I saw Rick was maybe a year and a half later in Miami, and no he didn't talk about Angola — but that was Rick. The other thing I remember when I first saw him back in the States was he looked different."

"How so?"

"Just a different vibe; he had a malaise about him. Let me put it this way — I knew two different Ricks. Okay? There was the guy I knew from the gun store, and he was a fun guy and he knew his shit, but he was kinda reserved. Now that guy in Africa was a totally different guy. In the jungle and in combat leading men is where Rick belonged. It was what he was put on this Earth for. His eyes would blaze, his focus was laser sharp, and he had a whole different vibe. You know?"

"Can you describe that vibe?"

"Killer, he was a stoned cold killer. See, that guy in the world was a part of Rick, a part of that renaissance guy I told you about, but Rick wasn't whole until he was in the jungle. Does that make any sense? See, to be honest with you all those stories and things you heard about Rick sounded good and cool, but maybe they were just stories. You know how people talk? So, you never know how a man really is or was in combat when you're just shooting the shit in a coffee shop, but Rick was all that and more. It's kinda like seeing a wild animal at the zoo. Their eyes are kinda dull and they just don't seem as dangerous, but if you'd encounter them in their environment it would be a whole different ballgame, buddy. You could bet on that."

"What else would you like to know?"

"How about Richard working for Bushmaster rifles? Do you know about that?"

"I know he was a salesman for them, and a damn good salesman at that, but I never directly asked Rick who his customers were. You just didn't ask questions like that. But he had a lot of clients always calling him, and he would take the big clients out to the Everglades to do demonstrations. Sometimes I would go with them.

"He had these Venezuelan military officers he would do business with, and they would always stay in those fancy hotels in Miami Beach. Rick back then had this big ugly orange Plymouth Trail Duster truck, and we would go by and pick them up. An hour and a half later we'd be in this one spot in the Everglades off of Tamiami Trail, just west of the turnpike.

"Let me try to give you a little visual of the area back then. Now remember I've been to some pretty dangerous and shady areas around the world, but this place took the cake. Once you passed this one last holdout of civilization — which was a little gas station off of Krome Avenue — you would enter this lawless dark apocalyptic world. Okay? Rick had his own hidden area he liked to go to, but we had to first cross through the Badlands to get there. The Badlands was the nickname me and Rick gave this one place.

"You knew you were getting close to the Badlands because it had a bad funk about it, a real putrid smell. It was a pungent mixture of rotting vegetation, spent gunpowder, and rotting carcasses. For some reason the catfish would leave the swampy water by the hundreds and slop into the mud of the Badlands, where they would either die or be shot. See, everyone went there to either shoot weapons or do some type of deals. There was never a time of day or night that you didn't hear some type of sporadic gunfire. The catfish weren't the only remains you'd find out there, because it was also the perfect place to dump the bodies during the cocaine wars. Every other week you'd read in the news that they recovered another body out there.

"Dig this. When I said it looked apocalyptic you got to picture that people would haul their old cars, trucks, refrigerators, washers and dryers out there to use for target practice. You'd see the bones of hundreds of rusted metal carcasses sprawled over dozens of acres, just riddled with bullet holes.

"To demonstrate for the Venezuelans Rick would bring out his unique weaponry. He would fire a regular looking AR-15 rifle about a hundred yards down range into a palm tree and shred it up pretty good. But then he would quickly swap out the bolt and barrel to create a new platform to fire this real heavy .450 caliber round. That round would hit those trees so hard they'd eventually fall. Most of the time Rick would have me do the shooting because he was the salesman. So, as I would be shooting the different rifles he would be explaining all the different benefits each one provided."

———————

"What do you know about Richard's drug arrest?" Sosa asked.

"Not much. The little I know comes from the court documents Dave emailed me. From what I remember reading Richard was arrested in late 1977 for eight counts of possession of large amounts of cocaine and marijuana with the intent to distribute."

"Well, don't believe everything you read. Let me start this next chapter in Rick's life with some info about the spook game. First of all, forget all that James Bond spy bullshit stuff you see in the movies. In the end what truly wins wars and overthrows governments is pure and simple intelligence, i.e. information gathering.

"It is no secret that the Green Berets have long been the military muscle behind The Agency. Okay? You see, The Agency — better known as the CIA — was prohibited from having its own operational forces, and realized they could best further their world-wide agenda by adopting the Green Berets. By doing so, the CIA managed to acquire operational capacity even beyond its most ambitious dreams. The CIA found its fertile recruiting grounds when the Special Forces established their home in Fort Bragg. There were also trainees from other countries in Bragg. The foreign trainees were supposed to be representing the military branches of their countries, but in all actuality they were hand-picked by their nations' intelligence organizations, and then they had to be vetted by The Agency. The Agency made it look like they were there attending a military aid program, but these men were really there for Special Forces training.

"You take most civilians and they have no clue of the difference between terms like a CIA agent and CIA officer. You see, the only people who actually work for The Agency are CIA officers, and it's only the CIA Operational Officers that recruit the human assets for the purpose of information gathering. These recruits — or subcontractors — are referred to as agents or operatives.

"Now, those agents and operatives sometimes do the same and recruit their own agents and operatives to work for them. So the layers get further from the actual CIA officer; this not only protects the officers with plausible deniability, but it allows the agents to operate in their own isolated cells.

"The real trick to field work is to one: make sure you're actually working for our side, two: make sure you're the one doing the playing and not getting played. Three — which is the most important — is staying alive. Now, this is how Richard told me it all went down with The Agency."

19

A Man Called Kates

1977 — South Miami
Norm's Hideout Lounge

Flaherty is at his usual spot at the local bar, sucking down beers as the jukebox beats out "Staying Alive" by the Bee Gees. Flaherty glances around the half-filled room and spots a young Cuban man playing darts with two beautiful girls. The trio is laughing it up, having a great time. In a rear corner of the business, sitting in a dark shadowy booth, is a well-dressed man in his mid-thirties. The man covertly watches Flaherty.

The Cuban man is showing off his dart throwing skills in front of the women by constantly hitting the bullseye from ten yards away. One of the girls spots Flaherty sitting at the bar with his legs dangling off the stool and nudges her friend. They stare and giggle at his expense.

Flaherty (as always) tries to ignore them. However, that timer in his head has already started its countdown. After several minutes of patience, the timer goes off and he hops off his stool, confidently approaching the group. The Latin man doesn't see Flaherty coming up behind him; when he does notice him he jumps in the air as if the circus has just arrived for his entertainment.

The man kneels down to Flaherty, like he's talking to a child.

Flaherty smoothly leans into the man. "You're pretty good with those. Any chance we could get a little action on a game?"

The man starts to chuckle and answers, "Are you challenging me, little man?" He turns to his two dates and says with mock gravity, "*Senor hombre pequeno con bolas grande* wants a little action."

He turns back to Flaherty. "What's your game, *Papo*?"

"One throw each. Whoever gets closest wins."

"One throw? What's the bet? Is twenty too much for you, big man?"

"You got any *juevos*? How about a real wager. You win I buy you guys drinks all night long."

"Sounds good. And if you should somehow win?" asks the man.

"Then your two lovely friends leave with me." The man and the girls burst out laughing. "Hey bro, my ladies are no *putas,* but I'll take your free drinks, you're on."

The man turns to the girls.

"You agree?"

The girls giggle and nod their heads yes. The man steps aside and motions for Flaherty to go first. With a half-bow, Flaherty backs away and says, "I insist amigo, you first."

The man, still laughing, grabs a dart and from over ten yards away throws a bullet, landing the dart very close to the bullseye circle.

"Little man, you better have the money on you because we're thirsty!"

The man tries to hand Flaherty a dart, but Flaherty refuses and stares at the target for several beats. The man thinks Flaherty is just playing, when all of a sudden in a blurring burst of speed Flaherty reaches below his pants leg to a worn leather scabbard concealed in his boot and pulls out his large Tru-Balance throwing knife. Without any wasted effort he underhandedly lifts his arm in a bowling motion towards the target and flicks his wrist upward right before the release to prevent the knife from spinning. It flies straight and true towards the target, impaling it with an audible *thud* just inside the bullseye circle.

The trio is stunned at the speed and accuracy they've just witnessed. Flaherty pays them no mind and walks forward to retrieve his knife. The man jovially starts clapping his hands and comes up behind Flaherty.

"*Hijo de gran puta*! Increi'ble! I don't know what circus you learned that trick at, little man," he gushes.

As Flaherty retrieves his knife the man continues talking, his hand falling roughly on Flaherty's shoulder. "But you ain't leaving here 'til you show me how ... "

Instinctively Flaherty grabs the man's arm and pulls it over his shoulder, using his own weight to judo throw him roughly onto the floor. Before the man can react Flaherty is straddling him, holding the knife to his throat.

The bartender deftly approaches Flaherty. "Hey Rick, he's good — he's had enough. C'mon man, he's just a stupid kid."

Flaherty pushes the blade one more time up to the Cuban man's skin, then slowly looks around the room at the crowd of stunned faces and gets off the man.

————————

Flaherty stares into the men's room mirror, tightly bracing his hands on the sink. He's waiting for the adrenaline rush to burn off, and for the shakes to stop. The shakes aren't because of the exertion, or how close he came to killing — the shakes only happen when he has to bring himself back from completing a kill. For men like Flaherty, once they pass the threshold of taking another man's life the line becomes as easy to cross as a crack in the sidewalk.

He splashes cool water on his face and starts the breathing exercises he mastered while training as a sniper. Inhaling through his nose and expanding his stomach for a count of four, holding the breath for a count of four, then

— he smoothly releases the breath, contracting his stomach for another count of four.

The restroom door slowly opens, and the man who was seated in the shadows walks in.

"My, my. That was quite a show out there, very impressive."

Flaherty doesn't acknowledge him, so the man moves to the sink next to Flaherty and starts to wash his hands.

The man slowly looks toward Flaherty. "A man could make quite a living performing feats like that."

"Show's over," Flaherty says in a flat tone. He turns and grabs some paper towels to dry his face. As Flaherty walks towards the door the man turns off the water and says in a flat tone mimicking Flaherty's, "I used to know a Montagnard in Laos who threw knives at dart boards, which of course is ironic because as you know the Montagnards don't have any sports or games with throwing involved, so ... "

"Who the fuck are you?" Flaherty cuts him off in mid-sentence.

"A friend who is going to help you get back in the game. That is of course, if you still want in?"

"I don't know what game you're talking about, but leave me out of it."

"C'mon Richard! Should a man who was capable of pulling off Operation Thunderbolt in the Quang Tri province be relegated to performing nightclub tricks in dive bars? Talk about wasted potential."

The man walks past the now-silent Flaherty and starts to leave.

"The name is Kates, and if I got your attention, I'm sitting in the last booth."

———————————

As Fleetwood Mac's song "Go Your Own Way" blares

from the jukebox Flaherty slides into the booth across from Kates.

Kates pulls out a leather attaché case and places it on the table. "Let's get to business. Your country still needs you, and I'm here to offer you a contract if you want back in on the action."

"If you know so much about me then you know I've already been burned. I'm not looking for any action."

Kates pulls out a memo pad from his case and scans it. Kates shakes his head side to side, very disappointed in Flaherty. "Let's see where you've been in the last few years. Hmmm. Angola, Botswana, and a little mercenary work in Rhodesia. Richard, that's the minor leagues for a man with your training. I'm talking about getting you back on team USA, working with The Agency again. We need to get you back to doing what you're really good at — kicking some commie ass."

Flaherty sits back and crosses his arms. "Need to do better than that, 'cause I certainly ain't going to work as a spook for The Agency."

"Not for us, but *with* us, let's say as a subcontractor. Oh, and I'll also throw in a Post Office job as your cover. As long as you're working for us it's a no-show job with a weekly check and full medical."

Flaherty leans forward, all business now, and quietly asks, "What do you need?"

"You familiar with the Nicaraguan revolution?" Flaherty nods his head yes.

"Carter is about to announce a 360 on current US policy; he's pulling all support for the Somozan government. We predict within the next year-to-sixteen-months that vacuum will allow the leftist Sandinistas to gain full control of the Nicaraguan government. Our plan is to start bringing in the most valuable players from the old regime and forming them up into little cells in Miami and San Francisco. Your objective is to provide the Miami east coast cell with whatever support they need to create a resistance army."

Coffee Shop Patio — Malibu

"Why would the CIA recruit Richard?" I asked Sosa.

"Let me first ask you this, Neil. Do you or your son know what Richard was doing with the 46th Company in Thailand?"

"We were told his assignment was to train the Royal Thai border police."

"And you and your son believed that? Do you really think the United States Army would waste highly trained Green Berets to teach security work to a bunch of Thai flunkies? Fuck no! The 46th Special Forces Company was created as a launching point for Green Beret teams to secretly insert into North Vietnam.

"Rick was there working with Army Intelligence, known as MACV-SOG, and behind them all was the good ole' CIA. The training bullshit was just their cover because they were also crossing the border into Cambodia and Laos. Part of the time he would work with the Haamongs. Am I saying that right?"

"I think the 'H' is silent, but go ahead."

"Well, Rick claimed them Hmongs were about the best fighters he ever worked with, and coming from Rick that's high praise. You know? They were especially effective at going into enemy territory and working as target spotters for American bombs. They were fearless, and they'd go on missions with Rick deep into Northern Laos — and I mean the real dirty missions nobody wanted to do. Great mountain men. Rick helped get one of those Hmongs out of country after the war, and they were good friends.

"It broke Rick's heart to see all those men who were recruited by The Agency to fight on our behalf abandoned in Laos after the commies took over. Many of them fled to Thailand to escape prosecution, but a lot of them were executed. Man, when you think about how many times in Rick's life he saw the American government abandon its own soldiers and allies you can understand why he eventually went a little crazy. And what about all those lies? You remember at that time our government was telling us that we would never go into Cambodia and blah blah blah.

"Rick's team also worked under Operation Strata — which I believe stood for Short-Term Road-Watch and Target Acquisition — on what was known as snatch and grab missions. Their targets were mostly comprised of North Vietnamese officers. And if Rick's team couldn't capture them then they were ordered to terminate them with extreme prejudice. So, you see going all the way back to 1969 the CIA was already fully aware of Rick and his capabilities. I'm sure you're familiar with the Iran-Contra scandal."

"Sure."

"Well, back when Rick got involved they weren't even called Contras — they were calling themselves anti-Sandinistas. Communism was just starting to take hold in Latin America and it was the CIA's job to stop that disease from spreading. The Agency's first goal was to help the anti-Sandinistas find an independent funding network that was completely untraceable. They also needed a form of currency that was accepted worldwide. The only currency the Nicaraguans could get their hands on for quick money was cocaine, and that was provided by the Colombians, who in turn needed distributors in the States."

"How could a man like Richard be involved in drug running?"

"C'mon, Neil, don't be so naïve. America has a long and dark history of funding wars with contraband. Rick knew of similar operations run out of the Golden Triangle of

South East Asia in 1969. He would use those operations as his template, however instead of heroin Rick would use cocaine as the currency. You know there's only one rule in war, right? It's to win at all costs."

"Yeah, but ... "

"But nothing. You know how many times Rick told me about the horrors of war, of him standing in pools of blood and going into the jungle to retrieve parts of his friends that were so small they fit into plastic bags?

"If Rick had to help move some cocaine to prevent America from sending our servicemen to Latin America to die in those jungles, he damn well would do it a thousand times over. Let the Nicaraguans fight and die for their own country, he would always say.

"Remember, this was way before the days of the crack epidemic. Back in the late seventies cocaine was a luxury item that only the rich and famous snorted in Studio 54 with their silver spoons.

"How'd he get the drugs in? With innovative thinking. At that time, if an airplane coming up from South America deviated its flight plan in the least bit, or dropped below an altitude, they would scramble jets out of Homestead Air Force Base. You know what a HALO is?"

"Yeah, a High-Altitude Low Open parachute jump."

"Exactly. Rick could carry seven keys on him — that was his max-out for the parachute rig he used, and that's a hell of a lot of coke. You're looking at what, roughly two hundred ounces, that's a lot of powder, man. He would pick up the product mostly out of Costa Rica from the Colombians and then he would duct tape the bags evenly to both of his legs, then set up a flight plan.

"He'd have the pilots fly out over the Everglades, and I had the keys to open some secured government gates that led me onto the hidden dirt roads out there. I'd just bought this big, badass 600 horsepower Ford 4X4 that towed a small swamp buggy, for just-in-cases. Remember there were no GPSs in those days, or anything, right? It's pitch black out there, I'm talkin' no ambient light 'cause we never ran an op on full moon nights.

"It would be me and sometimes another guy that I'd rather not mention. You'd hear the airplane once in a while, but most often you didn't. You'd just be sitting there, and all the sudden you'd hear, WHUMP! that sound a parachute opening makes. The next thing you would hear is Rick going, "Hey, how you doing," from behind you, and I'd literally come up off the ground. I'd yell, "Goddammit, don't do that!" and he would laugh his ass off 'cause he got a rise out of spooking me.

"But that's innovative there, buddy. That's how he got away with it for so long. There was just no way to tell. It was on one of those runs out of the Everglades that Rick told me fate stepped in and changed his life."

"The big drug arrest?"

"No man, much worse. He met the woman he would love forever!"

20

Swamps and Gators

0500 Hours
30,000 feet above the Everglades

Flaherty exits the plane into the freezing blackness of high altitude, breathing through an oxygen mask due to the thinness of the air. He checks his altimeter one last time and pulls open his chute at the last possible moment (2,000 feet), proceeding to land on a small dry piece of swampland. Flaherty takes out his map and a red-capped flashlight, surveying his location.

Flaherty walks several miles through dense jungle and swamp to an empty pick-up truck on the edge of the 'glades. Just as he's about to make it to the truck he twists his ankle in a small unseen depression and curses under his breath.

Flaherty limps up to his pick-up truck, swearing loudly at his mistake of losing concentration in the last minute of the operation. *That'll teach you not to be smug*, he chides himself. Luckily it's Sosa's turn to make the run across Alligator Alley with the product — he can finally go home and get some much-needed rest.

He drives a couple of miles through the dense swamp to the meetup point and gruffly hands Sosa the rucksack full of cocaine through the driver's side window.

"Tough day at the office?" Sosa laughs.

"Just take your time getting across the Alley, and call me if there's a problem," Flaherty plainly says as he drives off.

Flaherty turns out of the Everglades, driving down the

now-familiar dirt road next to the old run-off canal. Up ahead he sees a small car left haphazardly on the side of the road with the driver's side door half-open. He reaches under the seat for his 9mm Beretta and slips it into his lap. He slows as he passes, cautiously looking in all directions for threats, and spots a woman flailing in the middle of the muddy canal.

Flaherty pulls his truck to the roadside and re-scans the area, making sure this isn't some type of ambush rip-off. Satisfied that all's clear, he jumps out of the truck without thinking and painfully lands on his now-swollen ankle. With a noticeable limp he heads toward the canal for a better look.

He sees that the woman is trying to swim to the bank while holding a medium-sized dog in her arms. Although she seems to be a strong swimmer, she's having trouble keeping her head above water due to the dog's terrified flailing.

Without hesitation Flaherty jumps into the canal and swims out to her. He gets positioned behind her, throws an arm over her chest and pulls them both back to the safety of the bank. As she gets her feet onto solid ground he notices three things: she also has a limp, she's just slightly taller than him, and — she's drop dead gorgeous.

Flaherty assists her back to his truck as best he can on his bad ankle, the still-terrified dog clutched in her grip. He puts her in the passenger seat. She's barefoot, and has a sizable meaty gash on the bottom of one foot.

"Just stay put, I have an emergency kit in the back. That foot is bleeding really bad."

"I must have cut it when I first ran into the canal."

Flaherty — emergency kit in hand — expertly attends to her wound.

"There's all types of garbage dumped in those canals. Why were you and your dog in there in the first place?"

"Oh, him?" She looks down at the dog now lying by her feet. "No, he's not my dog. He ran across the road and I almost hit him. I guess he was so scared he jumped into the canal. He didn't make it more than five feet when he started sinking, and his head went under. I guess he's not much of a swimmer."

"Too much muscle and not enough fat to be a swimmer."

Flaherty laughs to himself, remembering Special Forces school. He had a few rough goes in the water, having to be revived a couple of times by the instructors. *Negative buoyancy is what they said I had.*

"By the way, I'm Lisa Davis. Thank you for saving my life."

Flaherty's Coral Rock House
South Miami

A warm sun rises. Flaherty and Davis are seated on his front porch. Davis has her bandaged foot up resting on the rail next to Flaherty's ankle, which is wrapped tightly with two bags of ice. The dog they rescued is sleeping under their feet.

Flaherty hands a rock glass splashed full of exotic colors to Davis as he holds his own glass in his other hand.

"I call this one the Cambodian necktie. It's six ounces of vodka, four ounces of lychee juice, and two drabs of Louisiana hot sauce."

"To your health," Davis adds, toasting Flaherty.

With his most winningest of smiles Flaherty answers, "L'Chaim!"

They both burst out laughing, and take long pulls from their drinks.

"So what's your tale, Cinderella?"

"You know, just the same old same old. A boring small-town girl story."

"Oh please, Ms. Davis, never bullshit a bullshitter. The necklace you're wearing is worn by desert Bedouins;

Eastern Sahara is my guess. Hardly fitting for a 'small-town girl.'"

"Western Sahara — and very impressive, Mr. Flaherty. What else do you know?"

"Please, call me Rick. Your accent isn't originally American. I hear the possible influence of British, with a slight mix of French?"

"Who are you?" she demands, smiling with amazement.

"Just a humble traveling salesman who happened to be in the right place at the right time."

They clink their glasses together and smile.

"Now who is bullshitting whom? I spotted military before you even made it to the canal."

"Guilty as charged, but let's first hear about you and your exploits."

"I was born in Morocco, Africa … Casablanca to be exact."

"Casablanca?"

"Yes, like the movie."

"Wait, wait. I have to say it after all these years of being called Rick!"

They look at each other, smiling, and then burst out together: "Louis, I think this is the beginning of a beautiful friendship." They clink their glasses again and laugh.

"My father was an Army captain, and I spent my childhood traveling around the world living on different bases learning about different cultures. By my teens my dad was stationed back in the U.S. I moved here two years ago to study biology at the University of Miami."

Flaherty is barely hearing her words. An electricity takes hold of his body as he stares at Davis, smitten by love at first sight. She keeps telling him more about her life and he keeps falling deeper into her spell, knowing this is a once-in-a-lifetime moment.

"From that moment on those two were joined at the hip. They would be either snorkeling in the Keys, boating down the Miami River, or just spending their time on the beach. And remember those parties I told you he would throw at the house? They were some wild ones."

"How so?"

"They had the most eclectic group of people I've ever seen at a party. There would be artists painting pictures on the back deck while Special Forces-types would be swapping stories over beers in the living room, while these New Age guru-types would meditate in the garden chanting to the rocks. Sometimes they would have a local band just show up and jam out, or some hippie would read poetry while the stoners lit up.

"I can still remember Rick in his kitchen cooking gourmet meals for everyone — and you know, come to think of it he wouldn't even touch a drink, not a beer. Now when I think back upon it, those times had to be the happiest days of his life. Maybe even better than walking point in the jungles of 'Nam."

"What about the operation? How was it going?" I asked, trying to bring Sosa back to the purpose of the interview — filling in Richard's missing years.

"I'm glad you asked me that, because I almost forgot. Well, Rick got another old Vietnam vet paratrooper buddy to handle most of the jumps for him. He also started to slowly back his way out of planning the operations. I know he once threatened to quit the entire operation when he learned that some of the money he was raising was going to The Legion of September 15."

"Legion of...?"

"They were a fringe group of anti-communist extremists based out of Guatemala that would sometimes kill civilians to get their point across. You know, like planting bombs in civilian planes and shit like that. Basically terrorists. Okay? Well, neither Rick nor any of us was about that, and we did our best to control where the money actually went, but Rick's heart was no longer into it. Shit, we all understood and covered for him. He was also in love, and I guess for the first time in his life he had too much to lose."

"So how come they didn't get married?"

"The mission got compromised and Rick got arrested."

"How?"

"The same way these things usually happen. One of Rick's Nicaraguan contacts who was already living in the U.S. for the last year met a nice gringa and got married. The problem started when the new wife found out he already had a wife — and a couple of *ninos* — in Nicaragua. She freaked out and told her uncle, who was a Miami narcotics detective, about all of his clandestine meetings. Right after that the guy got popped and rolled over on Rick to save his sorry ass, and that was the end of the operation. Rick was arrested and charged with eight counts of possession of cocaine and marijuana with the intent to distribute.

"Initially he was denied a bond because the judge knew of his history and training and thought he would be a flight risk. We tried to keep the truth from Lisa and stuck to the story that Rick was out of the country on a secret mission, but she had her own sources and quickly found out he was sitting in Dade County Jail looking at thirty years or so for being a drug dealer.

"And that was just too much for her and she ended it with him. Anyways, Rick would never tell her the truth of what the drugs were for. Even if he did, would she believe him? And that was that. She refused to speak to him, took their dog, and disappeared from his life."

At that point Sosa got emotional, stood up, and walked away from the table. I thought I'd lost him, but when he sat back down he was rubbing his eyes.

"Damn contact lenses keep drying up in this wind."

I said nothing. I knew it wasn't the non-existent wind.

"So, what happened in his criminal case?" I prodded Sosa, getting him back to the story.

"Government interference."

"The Army?"

"No, no, bro. Government interference ... The Agency ... they stepped in. He goes to court — he's looking at thirty years. Then he goes to court again, and it's now down to seven years. I might be off a year or two on the dates. The third time he goes, everything gets dropped to six months and a work release status.

"Judge, this is agent blah blah blah and Flaherty is an integral part of their operation to protect our nation and so and so. Come on, you don't go from thirty years to work release on the same charge after you've already been convicted. That's not happening. If that's not obvious government involvement I don't know what is.

"Now if you think that Rick got away clean you got another thing coming, 'cause the Agency had more work for Flaherty."

21

Razor's Edge

Biscayne Bay, Miami Bridge

On an overcast September day, Flaherty is sitting next to Kates on a park bench overlooking the bay. The sound of boat motors drones in the background.

"I told you I'm done. This shit cost me everything, and on top of that I now have a criminal record. There's no ... "

"Richard, hold on a sec ... "

"There's nothing you have that I want," Flaherty flatly says.

"The operation was a success. We helped build a resistance that actually has a chance. You did your job, you kept your mouth shut, and now I can give you what you always wanted. Just listen to the mission first, and then if you want to walk no hard feelings."

"You got one minute," Flaherty answers impatiently.

"Intel is coming out of Havana that there's dissension in the streets. An uprising could spark up any day now."

"Give me a break. As I recall things didn't go so well in sixty-one at the Bay of Pigs."

"Times are changing, and Cuba will always be a major threat to the U.S. until we shut down Castro's regime. Were you briefed on Brigade 2506?"

"Yes, and Alpha 66 and all the other splinter groups that still think they can take out Fidel. What's your point?"

Kates hands Flaherty a manila envelope and says, "Inside is a map of our current base camp in the Everglades — we shift its location every few weeks. You'll be in command of a squad of hand-selected men. You'll provide them with weapons and training. They'll maintain a high alert status, and must be ready to be transported within twelve hours of notice. Your final assignment will be provided on the day you and your team are activated."

"And the general mission parameters?" Flaherty asks.

"Covert assassinations of high-value targets inside Havana."

"Sounds lovely. Now, good luck finding another patsy to do your dirty work," Flaherty says. He stands up and starts to walk away.

"I'll do no such thing because I already have my man standing right in front of me. Like I said, I'll give you what you always wanted."

"Bullshit. What are you going to do, get me a job at the Coast Guard?" Flaherty asks sarcastically.

"No. I had you reinstated into the United States Army as a Special Forces reserve captain, stationed out of Fort Bragg."

Flaherty starts to chuckle. "Impossible. The Army wouldn't renew my contract back then; now that I'm a convicted felon there's no magic even The Agency could work to make that happen."

Kates hands Flaherty another file and says, "Then call me Houdini, because it's already done. You have your orders to report to Bragg on the first day of next month."

Flaherty looks through the paperwork in shock: it all looks legitimate.

"Oh, by the way, there's one last little thing I need you to do for me at Bragg. There's a couple of Green Berets inside the base that also worked in Central America doing some black bag operations. Rumor is they've gone rogue and decided to work their own operation. We have an Army corporal that got jammed up on a drug arrest and he's been feeding us info."

"So, you want me to make contact with him at Bragg and find out ... "

"Too late for that. Shortly after providing us with that info he accidentally drowned in a hotel bathtub."

"Drowning in a bathtub? Back in Laos, we saw a lot of that."

"How do you drown in a bathtub?"

"If you inject enough Ketamine into your target they'll fall into a catatonic state. You then place them in a bathtub, pool, or any body of water. The target can't stop himself from drowning. The real kicker is the Ketamine burns up in the body as he drowns, leaving no trace — so it always looks like an accident."

Kates thinks for a moment before answering, "The medical examiner's report showed only trace amounts of marijuana in his system, so who knows. What I want to know is was that corporal just blowing smoke up our ass trying to dodge drug charges, or is there really a viper's pit brewing at Fort Bragg?"

Coffee Shop Patio — Malibu

"Look, it's getting late and I got to start heading home. Is there anything else you want to know?"

I looked at my list of questions: they were all checked off. But with Richard Flaherty, one answer always leads to more questions. I asked, "If Richard was burned so badly by the military and CIA, why would he ever work for them again?"

"It's the juice, the high from walking on the razor's edge. Rick once told me when he was working missions he felt young again ... especially when he was out of the country. All his problems melted away and his nightmares would stop. These men are hooked on an adrenaline high that they can only reach when they're doing the most dangerous of missions. War is a drug stronger than any crack cocaine or heroin."

"How many different agencies do you think Richard worked for?"

"Rick was not biased. He'd work for almost anybody if he believed in their cause. As far as I know he worked for the Venezuelans. I know he's worked for several African potentates. I know he's worked for himself. And I certainly know he's worked for The Agency."

We stood and shook hands. Sosa started to walk to his car. Then, he turned around and came back to the table.

"Last thing, Neil. Please make sure you don't use my real name with this or anything else that can come back to me. One of my bigger fears ... okay, look. You want the honest-to-God's truth? I don't have that many years left, two maybe. If I do five, it's a god damned miracle.

"I've had serious health issues over the last couple of years. In fact, my oncologist can't believe I'm still walking and doing the things I'm doing. They all thought I was going to be dead a long time ago. I just don't need the hassle at this point with my name getting out there and people asking me more questions. I've got to get my wife to Tennessee where her sister is. Our older boy, who's also got some serious health issues, is coming with us. I just want enough time that I can take care of the wife and my boy."

Nine months after this interview Frank Sosa unfortunately lost his battle with cancer.

22

Undercover

ATF Special Agent Fred Gleffe, always helpful to my investigation, reached out to the ATF agent that first dealt with Flaherty and arranged a phone call between us. That agent, also recently retired, agreed to the call: however, he wanted to remain anonymous. I agreed, and Dayton is his pseudonym.

Agent Dayton explained to me that part of his assigned duties back then were to investigate illegal weapons possession and sales in the eastern sectors of the Florida Everglades. Early on Dayton started hearing the name Richard Flaherty buzzing through his confidential informants as a man that operated in between law-abiding citizen and a rogue outlaw. By the late 1970s, Dayton was getting regular reports from his informants that Flaherty was providing weapons and training to right-wing anti-Castro guerillas somewhere in the Everglade swamps of Florida.

Starting in the 1960s, the United States began numerous covert campaigns to overthrow the newly created Communist regime in Cuba led by Fidel Castro. Operation 40 was one example of a CIA-sponsored counterintelligence group whose mission was to seize control of the Cuban government once Fidel Castro was assassinated or ousted. Many other similar CIA-backed groups (like Alpha 66, Omega 7, Cuban Power, and Brothers to the Rescue) were formed in the sixties and seventies to assist the movement in various supporting roles.

In my first phone call with agent Dayton he made his position very clear.

"My problem with dealing with Special Forces guys is they actually think they're special. Now I heard all the romantic tales about men like Flaherty, who were only doing their duty to fight communism by helping these poor fighters. Well, I say that's mostly bullshit. Could these men have actually believed it was their patriotic duty? Sure, they might have, but the reality was guys like Flaherty were mercenaries who were getting paid extremely well to arm and train teams of highly dangerous men under god-knows-whose authority.

"One man's freedom fighter is another man's terrorist. We're a nation of laws, and when we allow this type of illegal activity to happen on American soil we forget what our nation was built upon.

"I tried on many occasions, in a one-year period, to infiltrate and collect evidence on him and his suspicious activities in and around the Everglades. Each time I got close to nailing him my investigation was blown, and Flaherty would manage to slip away. You see Flaherty's operation had the cooperation of several local law enforcement officers, and I always suspected a Federal agent or two was also involved in tipping him off.

"I needed to shut Flaherty down, and I finally found my chance to grab him at a gun convention in Fort Lauderdale. Even though I knew it wouldn't be the strongest case, it would be enough to get him in the system — and hopefully rattle him into cooperating with my investigation. What I really needed to know was who the top players were, and who was funding these operations."

I interrupted, "Was that the case with the gun silencers?"

"I guess he already told you about them, and I'm sure he claimed as a licensed armorer he was allowed to legally possess them, but I wasn't detaining him on possession. I was charging him with possession with the intent to distribute. I had informants that would testify he tried to sell them those silencers inside the convention.

"Well, Flaherty ended up being a tougher nut to crack than I originally thought. Even with federal charges looming over his head he refused to cooperate, or even work with me as a confidential informant. One day I was in Fred Gleffe's office talking to the guys about several of my cases when I brought up this stubborn little pain in the ass Green Beret Richard Flaherty.

"Well, Fred pipes in that he used to be a former Army CID intelligence man and maybe he could try to see if he could build a rapport with him. I told Fred he would have a better chance of getting blood from a stone, but he certainly could go ahead and try to work with Flaherty — even though I knew the man was utterly useless.

"The next day I called Flaherty to my office and he came in a couple of hours later. To shake him up I immediately handcuffed him and perp-walked him into the interview room. I then gave Fred a call to confirm I had Flaherty in custody and that he could come to the office and give it a shot."

ATF Interview Room

Gleffe walks into the small interview room and sits down opposite Flaherty.

"Captain Flaherty, how are you sir? I'm ATF agent Fred Gleffe. I guess things haven't been going too well with you and agent Dayton."

"So it's captain now?"

"Yes, sir. I'm a former Army man myself, and I would always give you that respect. I also took the time to read your military jacket. Very impressive, sir. This country owes you a hell of a lot — and I'm damn sure it's probably let you down more times than you can count."

Flaherty smiles to himself and says, "The world breaks everyone and afterward some are stronger at ... "

". . . the broken places," Gleffe finishes.

Flaherty perks up, suddenly interested. "I see I'm sitting with a cultured man. Finally someone in intelligence that actually has some intelligence! So, Agent Gleffe ... what can I do for you?"

"Well, I would like to help you, but you know we need you to cooperate first. Help us figure out who's really behind your operation. You scratch my back and I'll scratch yours."

Flaherty rubs the back of his head in thought and asks, "What unit were you assigned to in the Army?"

"Criminal Investigation Division."

"A CID man? Where?" Flaherty eagerly asks.

"I was stationed out of Fort Bragg."

"Fort Bragg!" Flaherty excitedly slams his open hand on the table. A grin flashes across his face. *I finally found the right man to help me put those motherfuckers out of business!*

"Yes, sir," answers Gleffe warily.

Flaherty, all business now, pulls his chair closer to the table and leans forward, speaking in a lowered voice. "Let's cut the bullshit, since I'm sure you have some idea of what really goes on at Bragg. Do you want to talk about real weapons smuggling? Do you want to work a case that's actually vital to national security? Might even make you agent of the year. Hmm? I'm not talking about some bullshit petty silencers case — I'm talking about literal tons of ammo and explosives."

"Tons?"

"That's right, tons. Claymore mines, M-67 fragmentation grenades, TNT, C-4, hundreds of feet of detonation cord, all up for sale to the highest bidder."

"I'm listening, but this is hard to believe," a shocked Gleffe says.

"Glad I got your attention. First things first, I only work

with you. There will be no Daytons on this case. And you immediately take care of squashing this silencer nonsense."

"Captain, I would love to take you up on that — I really would. But I make it a rule never to lie or bullshit anyone, including my arrestees or C.I.s, and I won't start now. There's nothing on this Earth that can get you out of being prosecuted by Agent Dayton."

"Unfortunately, time is of the essence. I can't sit here bartering like a street vendor. Please hand me that pad and pen, and from now on call me Richard."

Gleffe hands Flaherty a writing pad and a pen.

"I have information that a group I'm currently tracking inside Fort Bragg will be selling one of these on the black market in the next few weeks — and they might be able to get their hands on more."

Flaherty, pen in hand, starts writing out the words: *Green Light Project stolen classified weapons.*

As Flaherty continues to write Gleffe repeats, "Richard, I'm sorry. There's nothing you can offer that will change your situation."

Flaherty slides the pad over to Gleffe. Gleffe leans forward and intently reads: *Green Light Project.*

Gleffe laughs nervously. "Impossible. How could they?"

"Get your boss in here and I'll explain everything."

"Richard, if this is some type of ploy I can't even imagine how much worse you would be making things for yourself."

"You have my word as an officer and a gentleman."

Gleffe rises and leaves the interview room. Stunned, he walks into Dayton's supervisor's office and closes the door.

Gleffe and SAC Supervisor John Martin are sitting with Flaherty. Martin has a pad and pen in front of him, ready to take notes. His face is set in a fierce scowl.

"Listen, Flaherty," he growls, "if you're bullshitting us about any of this information life as you know it is over."

Flaherty's eyes sparkle as he smiles, delighted by the SAC's consternation. After a pause he says calmly, "No need for threats, John."

The SAC bristles at the use of his first name, but says, "Okay, give me the Reader's Digest version first. Then we can break it down into more detail."

"I first met Special Forces Sergeant Keith Anderson while at Fort Bragg for a mandatory training exercise. He was familiar with my military dossier, and also knew of my background working with weapons manufacturers. He told me he and his partner were starting their own weapons corporation, called C-MAG.

"We met up later that week at the PX — that's Post Exchange for you civilian types — and ... "

"God damn it, I know what a PX is. Now cut the shit, Flaherty!" Martin barks. Gleffe removes his glasses and pinches the bridge of his nose, working in concentric circles towards his eyes. *No good deed goes unpunished!*

Flaherty jovially continues, "Like I was saying, we met at the PX. Over some beers we started to exchange information. I was interested in his work in Central America and what type of progress we were making, and he wanted to know what lessons we learned back in Thailand and Cambodia that he and his teams could apply out in the field. Those early conversations established a foundation of trust as we swapped info — quid pro quo.

"I told him I would be in touch with him, monitoring his business to see if it grew to the point that we could start working together. In those early phone calls I quickly learned they were taking short-cuts and not applying for proper certifications and permits. Within a short period of time he asked me if I had any buyers for large quantities of rifles and ammo. When he quoted me prices, I knew that this was all for the black market."

Gleffe interrupts, turning to Martin, "John, by no means is that uncommon. As far back as I can remember as a CID man at Bragg, weapons and ammo were constantly walking off the base to be sold in the local pawn shops. I've always wanted to get the ATF involved up there and do a real sting operation, but the quantities and types of weapons being stolen wouldn't be worth our time. The other obstacle was it would be almost impossible to get undercovers into the Special Forces world. You'd have better luck trying to infiltrate the mafia."

Martin stares hard at Flaherty. "And you think you could penetrate this group and bring Fred in with you?"

"Yeah. Without a doubt," Flaherty confidently replies.

"Okay, you have a deal. I'll start writing up the operational plan and get the ball rolling with our North Carolina office."

Martin stands up. "Fred, un-cuff him. I want you both on our jet tonight heading up to Fort Bragg."

Gleffe takes the handcuffs off of Flaherty, who rubs his wrists, pretending they were too tight.

"Okay, but I'm not going anywhere until I have my deal in writing," Flaherty says stubbornly.

"After the operation ends ... successfully, and not before," Martin pushes back.

Flaherty leans forward, puts his hands out to be handcuffed and says, "Okay John, then you can explain to the attorney general how you had a chance to recover stolen classified weapons — and blew it."

Martin, fists balled, takes a step towards Flaherty. "You little bastard. You think you're going to blackmail me?"

Flaherty doesn't answer, but gestures for Gleffe to put the handcuffs back on.

"Fred, put it in writing and then get him the fuck out of here!"

Martin quickly turns and heads for the door, knowing the threat of having classified weapons available on the black market would haunt him day and night. On his way out he slams the door shut, leaving Flaherty and Gleffe uncomfortably staring at each other. Finally Flaherty breaks the silence.

"Well, that went better than I expected."

For the next hour Flaherty and Gleffe work out the details of his contract, which includes the ATF advising the Post Office that Flaherty is excused from work for the next six months.

23

A Man Called Griff

Inside the ATF's private Gulfstream III jet, Flaherty and Gleffe complete the details of their undercover plan. Once they land, Flaherty will call Keith Anderson and explain to him that he's in the North Carolina area escorting a high-rolling Miami drug dealer named "Griff." Griff needs Flaherty's help tracking down a man who owes him a large sum of money. Flaherty will then suggest a meeting between the two due to Griff's interest in buying large amounts of weapons for the Colombian Cartels.

August 14, 1984
Fayetteville, North Carolina

Gleffe sits behind the wheel of a rented white Cadillac Eldorado Biarritz as Flaherty exits a telephone booth in front of a Denny's Restaurant. Flaherty hops in the passenger seat and turns to Gleffe. "It's all set for nine at Green Tree Motor Inn."

"Okay, my records division should have my alias added to all the national databases they could have access to. That's if they decide to check up on me, of course."

"Not if. Trust me, they will."

Flaherty leans down and pulls his right pants leg up to adjust the holster carrying his concealed .38 caliber revolver. Gleffe spots Flaherty's furtive movement, sees the gun and slams on the brakes.

"God damn it, Richard! I told you before we got on the plane you're not going to be carrying. I'll put a halt to this mission right now if you pull any more stunts."

"Relax, Fred, not a problem. Just give me the key and I'll lock it in the trunk. But I do want the spare key in case the shit hits the fan."

"Give me the gun and I'll put it in the trunk. If the shit hits the fan and I'm dead ... then you can use my gun."

Flaherty's jaw clenches as he reluctantly hands Gleffe his revolver. Gleffe, still muttering, exits the car with the gun and walks to the trunk. *For all I know, this whole thing is a set-up. Flaherty and Anderson are probably going to blow my freaking head off the minute they get their chance.*

Flaherty watches Gleffe as he closes the trunk. *That was the gun I wanted you to see, but I got another .38 strapped to my other ankle 'cause there's no way I'm going into that viper's pit without some protection. Think I'm trusting you to watch my back?*

Gleffe and Flaherty exit the front office of the Green Tree Motor Inn with the key to room 165. As they walk Flaherty inhales the familiar North Carolina aromas of fresh-cut grass and barbecue-smoked wood.

Inside the small musty-smelling hotel room, Gleffe stares at the lime green shag carpet and matching curtains. *Wow, great place to get killed.* He quickly surveys the bathroom and checks to make sure all the lights are working. *Could this room be any smaller? What the hell am I going to dive behind if the shooting starts?* Inside the bathroom Gleffe checks the small Nagra audio recording device taped up his right leg to make sure it's working.

Thirty minutes later there's a knock on the door. Flaherty opens it on the large silhouette of Keith Anderson, framed in the doorway.

"Captain Flaherty. How are you, sir?" Anderson politely asks with a strong southern drawl.

Flaherty strides into the middle of the room with an air of supremacy.

"Good, good. Everyone let's have a seat: we have work to do," Flaherty responds.

Gleffe sits at a small wooden table as Anderson sits across the room on the edge of the bed. Flaherty pulls up a chair and sits between the two men.

"Sergeant Anderson, this is Mr. Griff. My Miami contact."

Both men nod, acknowledging each other. Flaherty looks at Anderson and proceeds.

"As I said, Mr. Griff was looking for someone in South Florida to bring him up to speed on working with all types of ordnance, and I was recommended. You got to understand nowadays in Miami every Tom, Dick, and Harry is importing product from Colombia; however, Mr. Griff is the real deal. He runs a very professional organization. I thought we all might be able to benefit from the introductions."

Gleffe leans forward. "Richard, if you don't mind? Mr. Anderson, Miami is so saturated these days with powder it's killing the market. What we really need are weapons and explosives. I have some potential buyers in South America that are looking for arms, and we've already started setting up purchasing channels through Bulgaria."

Anderson coolly responds in his deep voice, "Well, as I'm sure the Captain has told you me and my partners have started building our own solid corporation up here. Now, the Captain just called me this morning with the idea of working together, so if you give me a little time to talk to my partners about this whole setup maybe we can further discuss our mutual interests — at a later date?"

"That's a shame. I was actually hoping to bring some samples back to Miami. I brought cash."

"Well, I wouldn't be able to help you out that quick. Maybe in a week or two we could set up another meeting and talk some more."

"Not a problem. Me and Richard will be in town for another few days until we find this prick who owes me money. If you change your mind and want to talk earlier just let the Captain know."

Anderson turns to Flaherty. "Cap, you need help tracking someone down? This is my town. Maybe I could offer a little assistance."

"I already know all the spots where he hangs out. In the morning we're going to pay him a little visit," Flaherty answers.

Gleffe inserts himself back into the conversation, saying, "I'm just trying to deliver a clear message about being late on payments."

Anderson smiles at Flaherty. "We know all about messages. Ain't that right, Cap?"

Flaherty nods with a knowing smile.

"Maybe I could help you out. Mr. Griff, what kind of guy are we talking about?"

"He used to be reliable, but lately he's been real shaky. Before he skipped out of Miami he tried to offer up his ten-year-old daughter as collateral."

"Really? His own daughter?" Anderson shakes his head in disgust.

"Look, Mr. Anderson, I'm strictly a businessman — I don't mess with families."

"That son of a bitch! His own daughter? Shit, I'll do a scumbag like that for almost nothing. I can do it messy, creepy crawler style, or nice and quiet with some special K."

Gleffe gives his best confused look to Flaherty, hoping he can entice Anderson to explain the terminology as his hidden recorder continues to tape. Anderson, now excited, walks over to the table and sits down across from Gleffe to explain.

"You see, a creepy crawler is someone who crawls under a car and plants a little C-4. The best place to put it is under the steering column. See, bullets don't scare ya, bullets are easy ... but you take a guy out in a fucking car and spread that car all over half a city block, bits hangin' all over the street signs — then you're definitely getting through to them."

Gleffe shakes his head at the thought of the carnage.

"And the Special K is Ketamine. It's a muscle relaxer you inject into a major muscle group that dissolves in the body. In thirty seconds they go to never-never land."

"I don't understand," says Gleffe, hoping to keep him talking.

"See, it's perfect for making it look like an accident. You can just dump them off a boat and they won't move. It'll look just like they drowned."

"I'll tell you what, Mr. Anderson. Let me see if he has a change of attitude after I get my hands on him tomorrow. If he doesn't get the message I think I'd like to go with the C-4."

"Outstanding."

"Miami rate on hiring a real professional for a clean job is twenty grand. Would that work?"

"Twenty gees? Hell yeah, we could make it work. I'll tell you what: let me see if I can scrounge up a few of those samples for you to take home. Let's say we meet back here at twenty-three-hundred hours."

Gleffe stands up and shakes Anderson's hand, saying, "That's perfect. We're gonna grab a bite to eat and meet you back here."

After filling up on burgers and coffee at Kagney's restaurant Gleffe and Flaherty drive back to the hotel. As Gleffe and Flaherty walk towards the hotel room Anderson covertly watches them from across the parking lot, sitting in his Volkswagen van.

Gleffe walks up to the hotel room door and sees that it's cracked open. He quietly leans towards the door and listens for any noise. After half a minute he slides out his 9mm Beretta, turns to Flaherty and whispers, "Wait here." Gleffe cautiously enters the room without turning on the light and scans the room. Eerie half-shadows dance on the walls, caused by the bouncing light of the hotel's large neon vacancy sign.

Unnoticed by Gleffe the small dark shape of Flaherty silently follows directly behind him. As Gleffe walks into the interior Flaherty grabs his shirt and roughly tugs him backward. Gleffe spins and points his gun directly into Flaherty's face. Flaherty coolly reacts by holding a finger to his lips, signaling Gleffe to be quiet.

Flaherty kneels in front of Gleffe and slowly crawls toward a clear filament fishing line running across the middle of the room at ankle height. Flaherty follows the line until it attaches to a hidden booby-trapped Claymore mine. As Flaherty inches closer to the explosive, he starts to laugh.

"What the fuck?" Gleffe whispers from his now-parched mouth.

"You can turn on the lights — it's just a joke. That's a training explosive. Just a loud bang and a little smoke."

"Looks more like a message than a joke."

"Maybe both, but these guys could take you out any six ways from Sunday if they really wanted to."

Flaherty expertly disarms the booby trap and walks out of the room. He spots Anderson sitting across the lot in the van; Anderson gives him a big smile. Flaherty waves him over and Anderson exits the van, carrying a big duffel bag full of explosives.

Kosher Kingdom Bench

Richard was really animated and upbeat as he recalled the details of the mission. I finally had a chance to see the side Richard showed when he was doing the work he was best suited for.

"We later found out when we went for dinner that Anderson did as I predicted and ran the name Fred Griff in the national data system. The background check was run from an FBI agent's password, so we knew right away Anderson had some pretty serious contacts.

"From the best of my memory, that first night in the hotel room Fred purchased about a hundred pounds of high explosives. You know, the regular kind: C-4, TNT, military dynamite. Maybe nine hundred feet of det cord, and a box of maybe two hundred electric blasting caps mixed with fuse ignitors. Anderson wanted roughly thirty-three hundred bucks for the whole thing, but we got him down to three grand.

"After the deal went down Anderson sprang a proposal that even I wasn't expecting. Instead of just paying him in cash he and his partners would also accept cocaine and marijuana as currency. He claimed they were looking to become the biggest narcotics distributors in the area. Fred was ecstatic with the idea because he knew the case was escalating at a quick pace.

"We met up with Anderson a couple more times that week, building his confidence. Finally he invited us to meet his partners and see some of the inventory at their warehouse."

"Rich, what about the classified weapons — the Green Light project?"

"I'm getting to that, my boy, be patient."

Fayetteville, North Carolina
Shell Gas Station

Gleffe and Flaherty pull their rental car into a gas station as the morning sun starts to rise, promising another scorching hot summer day. Flaherty points to Anderson's van, which is already there and backed into a corner. Both vehicles then pull forward so the driver's side windows are adjacent to each other.

"Morning, Cap. Mr. Griff, if you'd just follow me it's only another ten klicks down the road."

Gleffe follows Anderson down an empty rural highway as they head deeper into the country. Anderson pulls off the highway onto a deserted dirt road, parking next to a steel warehouse.

As they exit their rental Gleffe sees several rough-looking men pile into a truck and drive off the property. Anderson motions for Gleffe and Flaherty to follow him.

As the men enter the structure they see an athletically-built man, with thinning gray hair and of medium height, holding a clipboard.

"Good morning, Captain," the man says with a heavy southern accent. He then reaches out and offers his hand to Gleffe.

"Byron Carlisle."

"It's a pleasure. Name's Fred, but everyone calls me Griff. "

"Mr. Griff, welcome to our corporation. Let me show you around."

As Carlisle walks into the interior Gleffe sees metal shelves piled to the roof with explosives, ammunition boxes, and wooden crates with military insignia, alongside VCRs and stereos.

"Looks like you gentlemen have everything but the kitchen sink," Flaherty says.

Anderson jumps in to add in a serious tone, "Military sinks and faucets are next to the 5.56 tracer rounds in aisle four."

"Excellent," Flaherty exclaims.

Carlisle turns to Gleffe as they walk. "Mr. Griff, if you're wondering why we're so free with showing you our inventory it's not just because you passed our background checks. It's because of the man standing next to you. Captain Flaherty vouched for you, and in the Special Forces world that man is highly regarded."

"Goes both ways, Mr. Carlisle. The Captain also vouched

for you. My people are very excited about doing business with you; we have the financial backing to take as much inventory off your hands as fast as you can stock it."

Carlisle and Anderson look at each other, trying to restrain their smiles.

"Well, we appreciate the business. If you'll follow Mr. Anderson into the office for payment we can start packing up your first order."

Warehouse Office

Gleffe and Flaherty are sitting at a desk across from Anderson as Gleffe counts out seven thousand dollars in cash. Carlisle is watching from ten feet away while leaning on the entrance door frame, smoking a cigarette.

A rusty fan whines in the background, attempting to cool the heavily sweating men. Anderson is openly carrying a 44-Magnum pistol on his right hip as he also counts the piles of hundred-dollar bills. One of the bills is blown off the desk by the fan and lands by Flaherty's feet. Flaherty leans down out of his chair to pick it up and notices that the long wire from Gleffe's hidden recorder is jutting out the bottom of his pants leg.

Flaherty picks up the bill, hands it to Gleffe and motions with his eyes for Gleffe to look down. Gleffe glances down and immediately sees the wire. Flaherty springs to his feet and stretches his back, turns and walks over to Carlisle.

"I can really use a cigarette."

Carlisle digs into his shirt pocket and retrieves a cigarette for Flaherty.

Sweat starts to pour off Gleffe's face as he casually leans down and tries to tuck the wire in before either man can notice.

Flaherty, in the corner, continues to engage Carlisle in conversation. Flaherty is subtly tugging at his pants

174

leg, pulling it slightly up in case he has to go for his gun. Gleffe finally finishes tucking in the wire and sits back up. Anderson stops counting and stares hard at Gleffe.

"Something wrong with you?" Anderson suspiciously asks.

Gleffe, looking pale, wipes the sweat from his eyes and answers, "Huh?"

Gleffe watches Anderson's hand slide slowly down to his holstered gun.

"I said ... is something wrong with you?

Flaherty — with his back turned to Gleffe — is carefully listening. He starts to bend towards his gun. Gleffe coughs several times and takes a deep breath. "Shit, I got to ease off a little with the nose candy. I've been getting fucking chest pains all week."

Anderson continues to stare hard at Gleffe as the silence becomes deafening.

"Jesus man, don't have a heart attack on me. We just got started doing business. Lay off that shit 'cause it will put you in an early grave."

Flaherty relaxes and goes back to his meaningless conversation with Carlisle.

Kosher Kingdom Bench

"After only six weeks of working the case we got word that the operation was being shut down," Flaherty said.

"Shut down? Why would they do that?"

"Fred didn't have to explain it to me; we both knew this was a lot bigger than anyone ever suspected. Anderson and Carlisle were low echelon, they didn't have the connections or resources to make this all happen. That amount of weaponry could only come from the top.

"One day while we were up in North Carolina Fred got the phone call. The word came from the head man himself, the acting director of the ATF. And by the way, Fred destroyed that hotel room. I knew he wasn't happy being pulled off the case.

"With the operation winding down we needed to recover as much as we could in this last deal, so we finally offered them the drugs they'd been asking for along with cash. It would be the biggest deal yet, over forty-seven-hundred pounds of explosives and ammunition for two kilos of cocaine and forty-nine-thousand in cash. The plan was for Carlisle and Anderson to drive down to Florida with the product in a rented U-Haul."

October 6, 1984
Days Inn Hotel
Vero Beach, Florida

At the hotel parking lot, Flaherty and Gleffe are waiting inside an immaculate Mercedes 300D sedan. A large dusty U-Haul truck driven by Anderson pulls into the lot on schedule. Flaherty hops out of the Mercedes carrying a briefcase and heads towards the truck. Flaherty enters the passenger side, puts the briefcase on his lap.

"How was the trip?"

"Smooth sailing all the way," Anderson replies.

"Where's Carlisle?"

"He had to take care of some things up north, so he's just going to wait for my call to make sure everything goes as planned."

"Not a problem. Let's get down to business."

Anderson pulls out an inventory list of everything inside the back of the truck and hands it to Flaherty. Flaherty quickly scans the list and nods: everything seems okay. He opens the briefcase, revealing it's full of cash and two plastic bags wrapped in brown plastic tape containing a powdery cocaine-like substance.

Flaherty tucks the list in his pocket. "Looks like everything we asked for is here. We'll unload the truck at the factory, then Griff will show you how to cut the kilos. Afterward you can take possession of this briefcase."

"Roger that, sir. I'll follow you guys over there."

Flaherty hands the two plastic bags to Anderson, who quickly stores them inside a gym bag. Flaherty takes the briefcase and jumps back into the Mercedes with Gleffe.

Gleffe pulls the Mercedes onto the road as Flaherty looks over his shoulder at Anderson's truck.

"Okay, he's behind us."

The two vehicles caravan down the road into an industrial section of town. After several miles Gleffe turns down an empty dirt-paved one-way street and stomps the gas pedal, quickly accelerating. Anderson watches the car pull away from him.

"What the fuck?" Anderson shouts.

Dozens of police cars swarm towards the truck as two unmarked federal cars block Anderson's path. Anderson slams on the brakes, bringing the truck to a halt. Gleffe swings his car in a U-turn and heads back towards the takedown area. Anderson throws the truck in reverse, spinning the tires. As the truck's tires gain traction on the dirt road they cause a massive cloud of dust to flood the air. Anderson looks at the side door mirror as he accelerates backward, seeing five more police cars now blocking his escape.

He slams on the brakes, causing even more dust to be thrown skyward, completely enveloping his truck. He instinctively reaches for the gun at his waist. Although he can't see through the cloud he hears dozens of cops and federal agents screaming at him to get out of the truck. Tension builds for the officers as they strain to see through the cloud. Flaherty darts out of the car, waving his arms at the officers.

"Gentleman, there's enough explosives in the back of that truck to kill everyone within a three-hundred-yard radius. Please let me handle this."

Flaherty jumps into a marked police vehicle and grabs

the car's microphone, switching the radio to the loud-speaker setting.

"Sergeant Anderson, we're standing down. I need you to abort the mission. I repeat abort the mission, that's an order."

A shape emerges from the truck as the dust cloud starts to dissipate. It's Anderson, walking towards the officers with his hands up. Officers rush forward, guns high, to take Anderson into custody; he offers no resistance. Gleffe and a couple of other agents rush to the rear of the truck and open the back gate. The inside of the truck is fully loaded with weapons and explosives. Gleffe smiles and receives pats on the back from the other agents. He walks over to Flaherty and shakes his hand. "Richard, you did an outstanding job. Thank you."

24

The Trial

On October 6, 1984, Green Beret Army Sergeants Keith Anderson and Byron Carlisle were arrested by Federal ATF agents on an 11-count indictment that charged them with conspiring to sell two and a half tons of mines, mortar shells, grenades, ammunition, and other arms to government agents posing as weapons dealers with links to drug smuggling.

On July 22, 1985, Anderson and Carlisle were tried in a Federal courthouse in West Palm Beach, Florida. The government's case was presented almost entirely through the testimony of Special Agent Fredrick L. Gleffe of the Bureau of Alcohol, Tobacco, and Firearms, and the hours of videotape and audiotape evidence collected during the investigation.

Federal Court House
West Palm Beach, Florida

Sergeant Anderson, dressed in his Army Class A uniform, is sitting on the witness stand answering questions from his defense attorney, Stephen Broudy.

"Sergeant Anderson, do you remember where and how you met Captain Richard Flaherty?"

Anderson answers in a deep voice with a strong southern drawl, "Yes sir, at Fort Bragg in ... in July of 1982. I believe it was sometime after the 4th that I met Captain

Flaherty during a training exercise. He was very well-respected in the Special Forces world, and after the training me and him got to do some talking."

"Can you give us an idea of what you discussed?" Broudy asks.

"Nothin' in general ... just our views of the world and politics. He was interested in my experiences in Central America and my thoughts on how we could get things turned around over there — to stop the spread of communism."

"And you had further discussions with him?"

"One night maybe a week later, I met him at the PX to have a couple of beers. That's when he started to explain to me what he called the big picture. He described himself as an international arms merchant.

"He went on to tell me that he was currently working with the CIA, and was also involved in covert operations down south, which usually meant somewhere in Central America. He didn't go into detail and I, of course, didn't ask. He was leaving Fort Bragg soon and told me he would keep in touch."

"And did he contact you?"

"Let's see ... Captain Flaherty might have called me eight or ten times over the next two years. Basically just checking in to see what I was up to. Then in August he called and told me that arms were needed for covert operations in El Salvador and Honduras. Because of political sensitivity such arms could, under no circumstances, be traced back to The Agency.

"The Captain told me about an individual whom he called "Griff," that he knew in Miami — an entrepreneur and drug merchant. Griff would be providing the funding for this venture; of course he would have no idea what the true purpose of the operation really was."

Broudy paces in front of the jury and asks, "Sergeant Anderson, earlier this week we heard audio tapes played by prosecutor Mancini of you inside several hotel rooms talking about some of the most heinous crimes that a man could commit. Now let me ask you, sir, was that actually you talking on those tapes? Could those tapes have been doctored up to sound like you, or chopped up in some way to change your words?"

"No sir. That was me, and that's what I said."

"Sergeant Anderson, I'm appalled. Let me get my notes to refresh my memory."

Broudy goes to the defendant's table, picks up a folder and walks back to the witness stand.

"You talked about bribing other Special Forces soldiers to obtain explosives. You talked about your master plan to create the biggest drug distribution network in Fort Bragg. You talked about using injectable drugs to kill a man, and you bragged about blowing people up in cars with C-4. Are you honestly saying that was you?!"

"Yes, sir."

At the prosecution table prosecutor Mancini whispers to his aide, "Here's the set up before he makes his client look like an angel. Wait for the pivot."

Broudy, in front of the jury with a concerned look on his face, asks, "Now, can you give the jury any possible reason whatsoever why you would say such detestable things?"

"Yes sir, I was puffin'."

"Puffing? Sergeant Anderson, what in the world is puffing?"

"Puffin' is looking bigger than you are. Flaherty needed me to convince these guys that I was the real deal. This guy Griff was supposedly some big-time drug dealer from Miami, and I've heard about what goes on down there ... "

A couple members of the West Palm Beach jury snicker at the Miami reference, happy that they're far enough away from those crazies. Broudy tries to conceal his smile — he knew that line would work. He listens as Anderson continues speaking.

"So, I figured I would come in there looking even bigger.

Acting like some gung-ho Green Beret character that's gone off the rails. Sir, if I may go on?" Anderson asks. Broudy nods his head approvingly.

"The whole point of the ruse was if we were ever detected, everyone involved in this venture would appear to be engaged in drug-related activity, rather than what it really was — a CIA-orchestrated covert operation. It's right out of the CIA playbook ... plausible deniability 101."

"Wasn't that an unusual way to get weapons to Central America?"

"No, because that's the way operations usually go down for Central America."

"And how were you initially recruited to go to Honduras in 1982?"

"Well sir, it was very similar to the operation with Captain Flaherty. I was invited to a clandestine meeting at the Bordeaux Motor Inn in Fayetteville by a man from.... A man from another government agency. I was asked a series of questions for about forty-five minutes that I'd rather not go into detail on. Next thing I knew I was told I'd been selected for a six-man team, and to report to another secret meeting at Special Operations Command. At the meeting they stressed how classified this operation was, and we were never to even mention the name La Vente. We were also told to grow our hair long and wear only civilian clothes — you know, jeans and T-shirts. We needed to be completely sterile: we had to leave our ID, dog tags, uniforms, and even our Berets in the States. We were then divided into two teams of three each and alternated going down to Honduras while the other team stayed at Bragg."

"Last question: why would you agree to volunteer for such dangerous missions? Like Honduras, and the one with Captain Flaherty."

"First of all, Captain Flaherty was a commanding officer. You don't question your superiors. The other answer is I love this country and would do anything I could to defend our great nation."

"And are you sure there was nothing else?"

"Well, to tell the truth Captain Flaherty mentioned that if I agreed to the operation I would also receive a promotion."

Federal Prosecutor Chris Mancini intently watches the jury's faces, hoping they aren't buying it. He waits until counselor Broudy sits down before he approaches Anderson.

"Good afternoon, Sergeant Anderson," he starts in a somewhat genial tone.

"In 1982, while at Fort Bragg, did you know what Captain Flaherty's status was?"

Anderson relaxes at Mancini's friendly tone and answers, "I believe he was a reserve back then, sir."

"And at that time, was he in the chain of your command? Was he your direct supervisor?"

"No, sir, not direct. But he was ... . "

"And in September of 1984, when Captain Flaherty called to recruit you to work this alleged covert CIA mission, do you know what his status in the Army was?"

"Well I'm not sure ... aah ... I mean, he might have been still active or something."

"Or something? Is that what you're saying, Sergeant Anderson?"

Mancini doesn't wait for an answer and forges forward; his voice is no longer friendly.

"You testified that you were just following orders from your direct supervisor, but that's not true ... is it?"

"Um ... well ... " Anderson hesitates, his body tensing up.

"In 1982 Captain Flaherty wasn't your supervisor. In 1984 he was no longer in the Army, yet you testified that you were only following orders. Isn't that right?"

"Well sir, let me ... " Anderson looks to his lawyer for help, but Broudy avoids eye contact.

"Aren't the rules and regulations plainly stated regarding Special Forces operators working with the CIA ... or any outside covert operations?"

When Anderson doesn't answer, Mancini continues, "Army regulations state that the Special Forces soldier will always first and foremost receive approval from the direct supervisor in their chain of command. What about that?"

"Well, there's by the book and there's what happens in real life," Anderson says defiantly.

Mancini walks up to Anderson. "Okay. Let's talk about real life. In real life, years before you ever agreed to work with Flaherty on covert weapons smuggling operations, you ... " Mancini levels his finger at Anderson and in his strongest tone yet says, "you were already stealing and stockpiling weapons, explosives, and anything else you could get your hands on in your warehouse for your alleged exporting. Now, isn't that right?"

Anderson sits stunned and stares at Mancini. He looks towards his defense counsel for help.

"Sergeant Anderson, do you want to explain to the jury how you were able to remove these weapons from Fort Bragg?" Mancini doesn't wait for a reply and rapid-fires another question to the now-stunned Anderson.

"Somebody must have been working with you. Was it blackmail or cash?"

The judge, frustrated by Anderson's reluctance, adds, "Sergeant Anderson, you must answer the question." Just like a boxer sensing a hurt fighter Mancini pours on the questions, not waiting for Anderson to respond to the judge.

"You never greased anybody's palm?" Mancini digs.

After a long pause Anderson quietly laments, "Yes."

"Oh, who did you bribe?"

"I'll take the Fifth,"

"Special Forces?"

"Yes."

Mancini hears the jury suck in a deep breath of astonishment. He knows he's just put the nail in the coffin.

"How far did this go?" the prosecutor asks.

"I'll take the Fifth," Anderson shakily repeats.

On July 29, 1985, after a one-week trial the case was submitted to the jury. Verdicts of guilty to all charges were returned for both defendants. Army Sergeants Keith Anderson and Byron Carlisle were each sentenced to incarceration for forty years and ordered to pay $13,076.13 restitution to the United States Army.

———————

One of the most uncomfortable phone calls I made on this investigation was when I cold-called Sergeant Byron Carlisle. I initially wanted to contact Keith Anderson first, but the records revealed he had already passed away.

I located a home number for Carlisle and dialed it with trepidation. How would he feel about a stranger bringing up his past, looking for answers about the man responsible for putting him in jail for so many years? As a cop, I always felt that once a man did his time and paid his debt to society he should be allowed to begin anew.

On the second ring a vibrant voice with a strong southern accent (I recognized it from an old news clip) answered. Fred Gleffe had sent me a taped interview from a 1984 news segment on the show *60 Minutes* that covered the entire case. Both Anderson and Carlisle provided their interviews in that clip from a holding facility.

I first uneasily introduced myself and explained why I was calling. I was met with silence. I forged ahead and explained from the beginning how I got involved in this whole matter. After another long pause, his first statement was as expected.

"You know that man ruined my life, ruined my military career, and had me put in jail for over twenty years."

Boy, where do I go from here? I thought, until Mr. Carlisle added, "Look, I was also in law enforcement. I was a State Trooper in Alabama, so what can I help you with?"

"Well, I always believe there are two sides to every story. I wanted to hear it directly from you and Anderson, not just from the newspaper articles and court documents that I read."

"So, I'm sure you read in our Supreme Court appeal that this was a CIA operation from beginning to end. I above all others should know best, because I also worked many times with them," Carlisle plainly stated.

"The CIA?" I asked.

"Yes, sir, The Agency. I'm not sure if you'll find any official record of it because I have nothing to do with that world anymore, but try to find information about a mission I worked in Honduras code-named Operation Quail Shooter."

"I'll certainly try."

"Well, let me try to fill you in. Back then in the late seventies and early eighties Central America was engulfed in communist flames, and the Reagan administration was on an ideological crusade to stamp it out. In El Salvador there were the Marxist guerrillas. In Nicaragua the new Sandinista regime was making sweeping revolutionary changes. Honduras, caught between the two, was being transformed into a pivotal American battle station against that red tide.

"The Honduran military at the time was run by an anti-communist general named Gustavo Alvarez — now remember that name, because it will become important later on. General Alvarez — our new ally with support from the CIA — formed a counterinsurgency brigade called Battalion 316 to root out Marxist infiltrators and weapons smuggling into his country.

"I was covertly sent to Honduras by the CIA with the military's full blessing to work with Alvarez and the Honduran military in Special Forces training. I'm not going to get into the specifics of what I did, but it wasn't any different than what Special Forces operators are still doing around the world today.

"You know there were many classified documents about those missions, and there were witnesses that the judge wouldn't allow us to call in the trial that would have cleared my name. You ever hear of the CIPA act?"

"No sir."

"It's the Classified Information and Procedures Act. It was allegedly put in place to stop the release of any classified information that the government felt could hurt national security. So here I am fighting for my career, fighting for my freedom, and the judge isn't allowing my attorney to bring up any information or witnesses that can exonerate me because of the CIPA law. My attorney Stephen Broudy attempted to subpoena forty-four different Army personnel, four CIA agents, and one FBI agent for my defense, and the judge ruled against us on all those witnesses.

"You look through all those court documents and tell me if you find the name Oscar A. Alvarez? You won't find anything, because his name was redacted. Remember I told you about General Gustavo Alvarez? Well, Oscar was his nephew. Sharp kid, and even went to college in the U.S. at Texas A&M. By the time I got to Honduras Oscar was already working with the CIA. I met Oscar at La Venta, which was the Honduran Army's Special Forces Camp located just north of the Honduran capital, Tegucigalpa.

"The CIA believed Oscar had great potential, so they sent him for further Special Forces training here in the U.S. at Fort Bragg. Now if this whole Flaherty thing wasn't a covert CIA op to get more weapons down to Central America, then what was Oscar Alvarez doing with me at that meeting with Flaherty and Gleffe? You talk to Oscar and he will clear my name. Also, after we talk you go look up Oscar and research what he did with himself. The kid had one hell of a military and political career. The reason Oscar couldn't be brought into our case was because the government didn't want the real truth to get out there, plain and simple.

"Now I also want to mention that it is possible Keith Anderson did things back then behind my back that he wasn't supposed to, but that has nothing to do with me. Last time I ever spoke to the man was when he called me about eight years ago out of the blue. Before he could say a thing I told him not to ever contact me again, and I hung up.

"Look, Dave, I wish I could help you more but I'm seventy-seven years old. I already lost too many years over this whole Flaherty thing, and I'd rather not give it any more of my time. Follow up on my leads on Quail Shooter and try to talk to Oscar. He'll clear a lot of things up for you."

My next call was back to my old friend Gleffe. He said, "Well, I also would have loved to talk to Oscar Alvarez, but he was completely hands-off. Look Dave, I can't go into too many details about this but I'll try to explain it as best I can. Obviously while working the undercover operation I had to provide constant update reports to my supervisors on what me and Flaherty were doing. But everything completely changed after I filed the report mentioning that Oscar Alvarez was at the meeting.

"After that I was ordered to secretly meet once a week with a liaison from another government entity that I will not name. Those meetings were on a need-to-know basis — Flaherty didn't have clearance for me to inform him, so he certainly never knew about it. It was only a week or two after meeting with the liaison that I was told to shut the operation down due to political sensitivity. I was also told in no uncertain terms that neither the U.S. nor the Honduran government would ever allow me to question Oscar Alvarez.

"Dave, I wish my bosses would have allowed the investigation to keep running and let me pursue the case further up the chain. In my opinion, Anderson and Carlisle were just middlemen. I know there were some really big fish involved at the top level that got away scot-free."

For the record, I had countless conversations with Fred and federal prosecutor Chris Mancini about the topic of Anderson and Carlisle being misled and therefore being innocent. Never once in any of those conversations did either of them ever slightly waiver in their opinion that both men were guilty as charged.

On March 24, 1985, *Washington Post* writers Joe Pichirallo and Edward Cody wrote an article titled, "U.S. Trains Antiterrorist CIA Military Foreign Squads." The article was quickly sealed by the CIA and finally declassified on February 20, 2013. The article talks about Operation Quail Shooter and mentions Byron Carlisle as one of the Green Berets providing training in the La Venta Special Forces camp. The article also includes mention of Carlisle's arrest in the Flaherty case. The article then takes an interesting turn by mentioning that Honduran Lt. Oscar Alvarez was at one of the meetings with Flaherty and Gleffe. Oscar Alvarez is quoted in the article as allegedly saying he knew Carlisle from La Vente, and was only at the meeting to discuss helping Carlisle with a plan to import wooden fixtures from Honduras. Alvarez declined answering any further questions about his knowledge of any secret training programs.

Reporter Frank Greve from the Philadelphia Inquirer newspaper also learned through a court affidavit filed in the case that Oscar Alvarez was named as attending one of those meetings. He attempted to publish an article on May 12, 1985 titled, "Cocaine Profits Linked to Latin Upheaval." That article was immediately classified by the CIA, with Oscar's name redacted from every official report. On September 26, 2012 that report was declassified via the Freedom of Information Act. The article mentions that Carlisle and Alvarez were friends and partners in a Honduran mahogany importing business.

My research on Oscar Alvarez reveals he was a highly decorated Honduran military officer that attended several U.S. military schools, including Ranger and Special Forces Training. He also received Special Recognition from the U.S. State Department, FBI, and DEA. After the military Alvarez became a successful Honduran politician and was named the minister of Public Security, putting him in charge of all internal security.

In his first term as Minister he played a key role in helping to implement Homeland Security's CSI (Container Security Initiative), which gave Hondurans the ability to ship containers to any port in the U.S. without major security restrictions.

On September 10, 2011, unexplained events led to the forced resignation of Alvarez and his closest staff. Since that time the sixty-two-year-old Alvarez has relocated to the United States. On March 6, 2018, he was named Director of Sunoco based out of Dallas, Texas.

It should be noted that on January 25, 1989, Oscar's uncle General Gustavo Alvarez was ambushed and killed while driving in Tegucigalpa, Honduras by seven men wearing the uniforms of a Honduran telephone company. He died in a vicious hail of automatic weapons fire from numerous Uzi machine guns.

Kosher Kingdom Bench

"How did you feel when you heard the judge's sentence? I mean, forty years for those guys is a lot of time," I asked.

Flaherty smiled at me and said, "Do you really believe that?"

"What do you mean?"

"Carlisle and Anderson were too valuable to the US

government to waste rotting in jail cells. I know for a fact that the CIA approached both of them with early release deals; in return, they would have to go back to Central America and do some more work for The Agency. Anderson agreed, and I believe he ended up only serving around three years. Then he was sent to a safe house, and from there who knows where he went and what he did. Carlisle, however, turned down the offer. I believe he might have served fifteen years or so in prison."

"I thought a safe house was only for when you were in trouble and needed to lay low?"

"Sometimes that's the case. Other times — for various reasons — they would place you in a safe house to isolate you for a few days or weeks. It's harder for people to track you if you disappear for a little while. I think Anderson was sent to the old Ledbury Lodge over in Elliot Key. Beautiful island and very isolated, but the god damn mosquitoes over there were almost as bad as the ones in 'Nam. It was a great safe house because it had a covered boat house attached to the main building. I know in the sixties a lot of top-echelon Cuban defectors were brought there."

"What about the Green Light Project and the classified weapons?"

"I would caution you by first saying it's probably not the best idea for you to know this stuff, but I'm sure you'd just investigate it on your own, so I might as well tell you.

"In the early fifties researchers at the Los Alamos nuclear weapons facility succeeded in miniaturizing atomic bombs from around ten thousand pounds to a much more compact warhead version — small enough to fit on top of a missile. Then, by the early sixties the newly created top-secret B-54 Special Atomic Demolition Munition entered the U.S. arsenal. It was only about two feet tall and was wrapped in an aluminum and fiberglass frame. According to an Army manual the weapon's maximum explosive yield was about one kiloton, which is equivalent to a thousand tons of TNT."

"What kind of damage could it do?" I asked.

"It would be powerful enough to take out the Hoover Dam or an entire New York City block, completely leveling all the buildings. Then of course you'd have to deal with the residual radiation, so you're talking about a highly effective weapon of war. Army Special Forces teams were trained under Project Green Light to carry the SADM by air, land, and sea to deliver them to enemy targets. They could parachute from cargo planes behind enemy lines, or swim with the bomb to its destination if necessary. Green Beret teams trained in special high-altitude parachute jumps were considered the best candidates to carry out the air missions. The SADM teams were broken up into two-man units and received weeks of specialized training at Fort Benning. They needed to be two-man teams because the two-man rule dictates no individual may have the ability to arm a nuclear weapon on his own. Therefore the Green Light team members divided the code that unlocked and armed the weapon between them.

"In 1968 after the Tet Offensive I heard rumor that a plan was conceived to end the Vietnam War in devastating fashion. Back then due to the lack of guided weaponry like we have now, dropping an atomic bomb on Hanoi would be immediately shot down because of the collateral damage to the civilian population — and of course the political fallout.

"So, the plan was for at least three separate Green Light teams to covertly insert into North Vietnam carrying the SADMs and plant them in and around secret enemy bunkers and installations in Hanoi, with the hope of destroying all of the North Vietnamese's top political and military leadership — including General Vo Nguyen Giap. Most intelligence agencies considered General Giap an even more important target then Ho Chi Min himself. So, if you kill Giap you're basically chopping the head off of the snake and ending the war.

"The project was very close to being authorized in 1969. However, the political experts eventually convinced the leadership that if we used nuclear weapons in 'Nam the Russians and Chinese would then fully commit, thereby kicking off World War Three. Also, the sensitivities amongst our NATO allies regarding these weapons were taken into consideration.

"By the mid-seventies rumors started spreading in the Special Forces world that one of those SADMs had been stolen. I only learned about its possible whereabouts a couple of weeks before I met Gleffe. Anderson told me that he knew who had the SADM and maybe he could set up a deal for it, but he also told me he was worried about it being out there. Even though Anderson was a stone-cold killer, he was still a patriot. He certainly didn't want the SADM to fall into enemy hands, so we started to work out a plan to recover it.

"Now, Dave, that's as far as I'll go talking about Project Green Light and the SADM. I won't get into how me and Fred recovered it, but I can assure you we did. It's now safe and secure on a secret military base — probably somewhere in Miesau, West Germany."

I knew Richard well enough not to push him any further for information. In all honesty, maybe he was right; I didn't need to know there were portable nuclear bombs floating around on this Earth. I had enough trouble sleeping at night as it was.

———————

Right after the trial Richard's cousin Donna Marlin told me she went down to Miami to visit him. She thought everything appeared normal at first, or at least as normal as things could be around Richard. However, she quickly noticed his alcohol consumption and paranoia were at an elevated level.

The first night she stayed in his apartment, Richard told

her he had to run out for a couple of errands. Before leaving Richard pulled Donna into the kitchen and pointed to an Uzi submachine gun that he'd bolted onto the kitchen counter, facing the front door. Richard gave Donna explicit orders to shoot anyone besides him who walked through that front door. He gave no further explanation of why, just told her to do it. Donna said she was so shaken up and nervous she couldn't sleep there, and went to a hotel instead.

After all the excitement of the undercover case, Richard helping the government recover weapons of mass destruction and making the world safe again — Richard Flaherty was once again forgotten.

25

Back to the Abyss

*"Back there I could fly a gunship, I could
drive a tank, I was in charge of million-dollar
equipment. Back here I can't even hold a job
parking cars!" —John Rambo, First Blood*

Kosher Kingdom Bench

Richard turned to me and said, "After the trial ... that's
when things started getting real dark. I was thirty-eight
years old and I wasn't where I wanted to be. I was still
working part-time for Bushmaster rifles, but business
was slow. I called Gleffe a bunch of times and told him
about some operations I thought we could work, but they
never seemed to pan out. I suspect that the ATF boss John
Martin had a lot to do with keeping me from working. I'm
sure Dayton also wasn't happy when Fred won Agent-of-
the-Year for that case and his name was never mentioned.

"I then got a call from the Post Office informing me
I must report for work now that I was cleared from all
my other governmental responsibilities. I'll tell you what
Dave, I did try to give it a go. I tried to make it work at the
Post Office, you know — work like a regular nine-to-five
guy. But things kept falling apart.

"Some of it was my fault, but you also have to remember I'd made powerful enemies over the years. Enemies

that would pursue me for however long it took to make sure I was finished."

"People trying to kill you?" I asked.

"No, no, for heaven's sake, no. Killing me would be too easy. No, they want to destroy me, make me look like I'm crazy. No, killing me would be far too generous of them. They want to make my life a living hell. You remember last year when I got beat up?"

"Yeah, it was supposedly a couple of teenagers."

"There were no teenagers. That was just something I made up in the hospital when they were bandaging up my head. The guys that roughed me up were from the State Department."

"State Department? I'm sorry Richard, but why would the State Department send guys to rough up a sixty-eight-year-old man?"

"To send a message. To let me know they're still tracking me. To make my life miserable. I know they're the ones who interfered at the Post Office when I tried to actually make it work."

"Okay ... but why?"

"Because I know too much. I know too many of their secrets. That's why they won't just make it look like I had an accident and kill me. They're worried I wrote everything down, maybe stored it in some safety deposit box only to be released to some Washington newspaper reporter when I'm found dead.

"You know, the pen is mightier than the sword-type of shit. No, no, my boy, they want me alive so they can watch me and keep discrediting me. That's why the VA hospital always denies my benefits."

Richard pointed off into the street, his speech pattern taking on a rapid-fire cadence. "Here is where they want me. That's the sick game I'm trapped in. This is my life. Don't you notice all the helicopters that are constantly flying over this city? Yeah yeah, some of them are routine, but every once in a while I'll hear a Huey. A Vietnam-era

Bell UH-1 helicopter. They do that to try and cause me to think I'm going crazy. Don't you get it?"

I tried to think of some response as I digested his words — but what was there to say? *"I'm sorry, Richard, but this conspiracy train ride is too much for me to go along with. I'm getting off at the next stop,"* I thought. Richard spared me by not asking me if I believed him or not. He continued with, "See, I didn't become homeless overnight. I slowly made the transition. And sure, part of it was financial. Without the Post Office checks coming in I couldn't keep up the rent at the Coral Rock house. I went from a smaller apartment to a smaller apartment, until one day I put all my stuff in the storage unit and that was it — I was free. I was off the grid. I would blend into the streets and make it harder for them to track me. I wasn't going to wait for my enemies to come to me any longer: I would start to hunt them."

Richard now appeared very agitated. I could see a visible change in his breathing as he furtively scratched the top of his head. For the first time he wore an expression of defiance as he sneered up at me. I tried one more time to calm him and get him back on track filling in that time period of the late eighties and nineties, but in a dismissive manner he waved his hand and stated, "There's really nothing too interesting to say about that time, nothing at all."

My research later revealed that by the late eighties Richard cut all ties to friends and family and started traveling around the country. He first traveled throughout the state of Florida, then headed up to Texas, North Dakota, and Alaska.

At first I figured he was just traveling to visit some of his old Vietnam buddies; I'd found paperwork indicating he requested their addresses from the military. But his trip to North Dakota (alongside some other documents I would later discover) raised red flags with me.

Documents dated in 1986 reveal that Richard was no

longer an employee of the United States Post Office. From 1986 to 1990 Richard stayed off the radar. There isn't anyone I've spoken to yet who knows of his activities.

In 1990 Richard was arrested for public intoxication in Lee County, which is located on the west coast of Florida near Fort Meyers. I was told by one of his old friends that in the early nineties he believed Richard spent a lot of his time in the Florida Keys, specifically Key West. Other records I found revealed that in 1991 Richard had no listed address, but he did have a P.O. mailing box address in Key West.

Kosher Kingdom Bench

I tried changing subjects with Richard, asking him about any future travel plans, but he ignored my question and continued to poke uncomfortably at the top of his head. Some of the medical tape on one of the gauze bandages had started to come loose. I headed over to the trunk of my car and pulled a roll of tape and gauze out of my medical kit. By the time I came back Richard was holding the old gauze bandage in his hand.

"You want to go to the hospital and have them replace it, or you want me to do it?"

Richard defiantly stuck his hands out, indicating that he'd do it himself.

"I got it," I said.

As I was replacing the one-inch square gauze I saw that he had some bleeding and skin loss on the top of his head. While I did my best impression of someone with medical training Richard looked up at me. "Don't worry, it's nothing you can catch. About ten years ago I had some sunburns that wouldn't heal, so they sent me to a dermatologist. After a scalp biopsy they diagnosed me with squamous cell carcinoma — better known as skin cancer."

"What kind of treatments are you getting?"

"Goddamn VA made it worse. I should have never trusted those bastards. They used some type of radiation therapy — they said it was low energy. X-ray beams on me only made it worse, I'm sure. You see, they'll never outright kill me ... no, they'll just make me suffer a little bit more."

"That sucks. Sorry to hear that."

Richard then became engaged in a private conversation, animatedly talking to himself. I couldn't hear what he was saying because he was mostly mumbling, so I quietly and uncomfortably just stood there trying to be patient. After a minute or so he snapped out of it and looked up at me, smiling.

"What did I tell you about bad things, hmmm? I told you that out of the bad something good always happens. Sometimes you need a little kick in the ass to get going. So, after hearing about my biopsy I decided it was time to finally look up an old friend and give her an explanation."

26

Beyond Love's Limit

"Of all the gin joints in all the towns in all the world ... —"Casablanca"

In the early 1900s Dr. David Fairchild was a scientist, a teacher, and an avid explorer who traveled the world in search of plants that could help humanity. He journeyed to almost every continent and brought back hundreds of species of plants, including mangoes, alfalfa, nectarines, dates, cotton, soybeans, bamboos, and the flowering cherry trees whose ancestors still grace American parks.

In 1935 he retired to Miami and joined a group of passionate plant collectors and horticulturists, including environmentalist Marjory Stoneman Douglas, Commissioner Charles Crandon, and landscape architect William Lyman Phillips. As a group they worked tirelessly to bring the idea of a one-of-a-kind botanic garden to life, and in 1938 Fairchild Tropical Botanic Garden opened its eighty-three acres to the public.

2005 — Fairchild Botanical Gardens
Miami, Florida

A young park employee dressed in beige khaki shorts and shirt escorts Richard Flaherty past the giant baobab tree at the entrance and into the lush botanical garden. Flaherty is neatly dressed, wearing a white collared dress shirt with crisply pleated slacks and mirror-polished black shoes.

The park employee nervously asks, "We don't have many Department of Agriculture agents down here. Are you sure you can't tell me what this is about?"

"Need to know basis only, sorry my boy," Flaherty flatly replies.

Forty-seven-year-old Lisa Davis is kneeling on the ground in one of the many gardens, pruning at some flowers. Surrounded by picturesque garden foliage and halos of colorful butterflies, Davis' natural beauty seamlessly melts into the landscape. Flaherty approaches Davis while she works and observes her for a few moments. She is engrossed in her work, still hasn't noticed him. She is the one woman he's loved for so many years, and he's afraid to break the serenity of the moment. *Will I ever know happiness again?*

Flaherty hesitantly clears his throat and interrupts the silence by saying, "You know, the Egyptians revered their flowers so much the Pharaohs would adorn their carts with water lilies before heading off to war. There were actually two types of water lilies that grew in the Nile: one was the blue lotus and the other was the Dayton lotus, or Nymphaea lotus."

Davis continues to prune, not looking at Flaherty, although a noticeable tension floods her muscles. Still not looking up she finally answers in a constricted tone, "Richard Flaherty. I swore if I ever saw you again the first thing I would do is slap your face."

Flaherty is stunned and at a loss for words — a rare occurrence.

"Well, hmm, let me see. Yes, other than color the two lilies differ from each other in several other attributes. The blue lotus has pointed flowers and floating leaves, while the Dayton lotus has rounded petals and ... "

Using her small pruning shears, she lops off the heads of several innocent flowers.

"You hurt me more than you'll ever know. It took me years to get my life back on track. Years! Why are you here?"

Flaherty takes a step closer to her and lowers his voice.

"Because I had to see you again. Because I can't live with myself knowing the pain I caused you. I always wanted a chance to see you in person, to apologize and to say ... "

Davis abruptly stands up, with her back still facing him.

"Get out of here, Richard. I don't want to hear another word — and I don't want to see your face."

Davis grabs a walking cane leaning on a nearby bench and attempts to walk away, but she loses her balance. Flaherty rushes forward, catching her. The two look into each other's eyes for the first time in almost twenty-five years. A long moment of silence passes; Davis finally exhales a deep breath of decades-long anger, placing her hand over Flaherty's.

"What's wrong? What happened to you?" Flaherty concernedly asks.

"It's a long story, but it started the day I met you. It's my heart."

"Me? Because I broke your heart?"

A big smile floods her face. Warmly she says, "Yes, you broke my heart, but that's not what I meant. Do you remember that cut I had on my foot when I came out of the canal?"

"Yes, of course. It would get red and swollen every once in a while."

"Well, they believe I must have stepped on a rusty piece of metal. The infection has been in my bloodstream all these years. It sometimes makes my heart work a little too fast, and it causes me some dizziness."

"What can I do? We need to get you to specialists. I know some people at Holy Cross Hospital that ... "

Davis lets out another long healing breath and adds, "I've seen the best doctors. My fiancée Larry has made sure of that. They believe my symptoms will improve with the new medication they've started me on. It's just going to take a little while to get the poison out of my system."

Flaherty carefully asks, "Are you happy?"

Davis avoids the question and slowly starts to walk forward, Flaherty supporting her arm.

"With him?" questions Flaherty.

"Yes. Yes, I'm very happy. Larry's a good man, a hard-working decent man. For years I refused to go out with him because he was also a Vietnam vet, and it was just too close to home for me. Damn can I pick them. I must have a magnet for your type."

"No, I don't think so. It's probably because you were always so special and beautiful, and still are. You need someone capable — worthy — of protecting you."

"Always the charmer, Richard J. Flaherty." She sighs and quickly becomes serious. "Are you still having the nightmares?"

"They come and go, but I think they're getting better."

"And what about you? I don't see a ring. Why isn't there a Mrs. Flaherty?"

"Look Lisa, I'm not here to try to win back something I lost. I know that ship has sailed, but those were the happiest times of my life. I just needed you to hear that. Also, I needed to know how your life was going — if you were happy. Now that I know, I am truly happy you have someone to take care of you. You deserve it."

Flaherty lets go of her hand and adds, "Lisa, it really was great seeing you. I promise I won't ever bother you again. I'm just going to say for the last time how truly sorry I am about not being honest, and for the pain I caused you. So that's that. Goodbye my love, and take care."

Flaherty turns and walks away. Davis tries to turn her head away from watching Flaherty leave, but can't.

"Richard, wait!" she yells, then pauses to think. "Why …

um? I'm sorry, you were saying about the Egyptians and their flowers."

Flaherty, stunned, stops walking and gathers his thoughts. He spins around and confidently strides over to her. "Well yes, as I was saying according to the myths it's believed that the ancient Egyptians used to sing for the lilies at their festivals."

Davis, cane in one hand, takes Flaherty's arm in the other. The two start to stroll down the lush garden path in the warm glow of afternoon light.

Flaherty continues, "During the festivals, everyone was supposed to bring a bowl with a candle burning in its center. Then they would walk down to the Nile River with the bowl in hand and an overwhelming dream in their heart. They would gently place the bowl in the river, and if the burning candle continued floating on the surface they believed their dream would come true."

For the next three months Flaherty, once a week, travels over four hours — transferring onto three different city buses — to visit Davis at her park. They stroll through the park's beautiful meadows, sit on benches, and have deep conversations.

As they sit by the pond, Flaherty takes the wooden bowl he always brings with him from Davis' hands and lights the candle inside of it. He places it on the pond and they watch quietly as it gently floats away.

Every week Davis is having more trouble walking. By the second month her cane has turned into a walker. The next time he visits Richard is pushing her in a wheelchair.

They stop at their favorite tree, which overlooks the pond. Flaherty places another candle-lit bowl in the pond and closes his eyes, praying for the Gods to intervene and cure Lisa as he gently releases it into the water.

When he turns back to her he sees several multi-colored butterflies flock into the area. One lands on the armrest of Lisa's chair, causing her to smile. Observing the butterfly she says, "If I could do it all again I would have become an entomologist and devoted my life to these beautiful creatures. Who would ever believe something that starts out so small and ordinary could become so incredible?"

Flaherty paces several steps and rubs his head in thought, trying to remember a poem he once read. Finally he replies,

> *Happiness is a butterfly,*
> *which when pursued,*
> *is always just beyond your grasp,*
> *but if you sit down quietly,*
> *it may sometimes come upon you.*

Davis melts with affection at Flaherty's depth and reaches for his hand. The two warmly hold hands as the breeze slightly rustles the leaves on the trees above them. Davis concentrates on her reply, then recites,

> *The butterfly counts not the months*
> *but the moments, and has time enough . . .*

Davis continues to grasp Flaherty's hand as the two gaze into each other's eyes.

Kosher Kingdom Bench

Richard stood slowly from the bench and faced away from me. He paced up and back, quietly muttering some long-ago conversation or argument. I could tell that he probably relived this moment in time, this one conversation, every day of his life.

For the first time I heard Richard's voice sound strained as he said, "The last time I went to the park to see her was Friday, April 29th. Her coworkers told me she died the day before in her bed."

There was nothing to say. I felt as if we were suddenly

at a high altitude — the air was harder to breathe, thinner. In a lifetime full of disappointment and sadness why did he have to lose her too?

All I could get out was, "I'm sorry."

Richard walked away for a few minutes and composed himself. In the world where Richard and I came from you didn't coddle another man when he was in pain, because that would only shame him further. I kept my face unreadable and waited until he sat down again, ready to talk. Then I played the role of the good listener.

"She was right about that fiancée of hers being a good man. A couple of months after she passed he bought up several acres of land at that park and renamed it after her. She would have loved that."

After a long uncomfortable pause, I tried, "You know, I'm off this weekend. Why don't we take a little road trip? I've never been to a biker show and they're having a big Harley-Davidson rally up in Daytona this weekend."

"I appreciate that, my boy, but I'm too old for that shit. And who would keep an eye on my tree? If I left town it could get broken into."

"I hear a lot of Vietnam vets still go up there with their bikes. Maybe you'll run into an old friend?"

"Yeah, I guess it's cool now to be a vet and ride your Harley with your military patches sewn onto your vest, but I remember a time when it wasn't like that. The country was a lot different back when we got home — all the hate. You know what I'm talking about? It's happening again right now in this country. It seems like they're killing cops every day, and I see people in the news talking about how fun it would be to shoot a cop."

He was right. It was happening. In almost fifteen years of police work I'd never experienced so much hatred from the public. There were people who moved away from me when I sat in restaurants, bystanders taunting and cursing at me even when I was handling routine calls for service, like car accidents. Wearing that uniform felt like a big target. Yeah, I could understand a little of what Richard was talking about.

Richard became very angry, clenching his fists. "They don't know what it's like being out in the jungle and eating bugs. They criticized us and criticized us ... " He paced back and forth like a caged tiger, muttering under his breath until he calmed down enough to sit back down.

"Was it worth it? The war?"

"I didn't see the war as the wrong thing to do at that time in history. I could go into all kinds of detail about us back then being part of SEATO, which is the Southeast Asia Treaty Organization, and our obligations to the other members — but it would be an exercise in futility. The way I see it is when your nation calls upon you and you've taken your oath you go.... It's that simple. You do your job.

"What pisses me off and disappointed me the most was how the politicians sold us out. We could've easily — no, I shouldn't say easily, that's wrong. But we could've won that war. We were too restricted from doing things that needed to be done."

"Fifty thousand men losing their lives while the politicians were stuck in indecision is tough to comprehend," I said.

"I agree."

"What's the one thing you remember most about Vietnam?"

"The smells. The jungle, the shit burning, but mostly the smell of blood. The smell of blood: it's coppery, and there were times that it really got to me. Especially after several of your own people have been killed ... you know it's never neat, it's messy. It's always messy, a lot worse than people see on television or in the movies. Just being here in Florida with the tropical smells can sometimes bring me back.

"Yeah, I mean even the heat and humidity is relatively the same. You go over there and it's hot and nasty all the time, and you're living like a wild animal. I used to tell people war is men hunting men with guns. It's just like wild animals going after each other, except you've all got guns.

"And you see good people get killed. They're there one moment, gone the next, and you don't ever forget them. Those men I lost are with me every day ... every day.

"I don't think you can come out of a combat situation and not be affected by it, especially when you're in the infantry. The infantry is up close and personal. We're not firing a 105 Howitzer from ten miles away, you know? It's the kind of fight you can't afford to lose. I'm thinking if you shot at somebody one hundred yards away — that was a rarity, an extreme rarity. Because almost everything was either twenty or thirty yards. There could also be hand to hand — it could sometimes get that bad.

"You know my boy, I wish I could tell you the betrayal by the politicians and coming home to a country that hated me was the worst part of the war, but there are things that are much worse."

Uncomfortably I asked, "Like what?"

"Guilt. The guilt of being in charge of a group of men and not bringing them all home. The guilt that I'm still alive and they're not. It's the heaviest weight a man can bear."

Richard got up and hoisted that heavy backpack onto his shoulders. I realized for the first time the backpack carried a lot more weight than mere material objects. Without saying another word he turned his back and walked away, already engaged in another ghost conversation, trying to sort something that could never be undone.

27

Nightmares of the War

"Those who die perish in excruciation, but those who escape are perpetually scarred with the memories of such pandemonium and horror. War shapes and haunts the soldier's eternity." —Oliver Stone, "Platoon"

Publix Supermarket Aventura, FL
December 5, 2012 — 9:30 P.M.

Alone and shivering under his palm tree, Flaherty struggles to stay warm despite wearing a jacket, snug knitted cap, and gloves. It is the first cold front to hit Miami this year, and the forty-eight-degree weather causes Flaherty to shake uncontrollably. Flaherty, always hating the cold, decides to grab a warm meal in the nearby Publix supermarket before the store closes for the night.

Inside the Publix Supermarket Marine Corps veteran (now seafood manager) Howard Singer is just starting the nightly closing ritual of putting the seafood from his window display back into the subzero freezers.

Flaherty enters the business and heads down an aisle. He pulls a Hungry Man frozen food entrée out of the refrigerator and brings it over to the deli section, hoping they will warm it up in their microwave. Howard Singer, after finishing his responsibilities for the night, starts walking out of the store when he hears a loud commotion by the deli.

Singer walks over to see a female employee — in the process of closing up her section — yelling loudly at Flaherty. "I told you I can't warm up stuff for you anymore. Last time I did I almost got fired." Flaherty, still holding out the entrée, doesn't respond. Instead he mutters softly under his breath and turns to walk back to the frozen food section.

Singer walks by Flaherty and whispers in his strong New York accent, "Come with me, Rich. I'll take care of you bro." Flaherty follows Singer back to the seafood department. Singer discretely grabs Flaherty's dinner and brings it into the back to warm it up.

Publix night manager Mike Thornburg is completing his monthly inspection of all the stations when he notices Flaherty standing in front of the closed seafood section. Thornburg — with his muscular frame, shaved head, and serious demeanor — looks more like a Navy Seal than a supermarket manager. Just as he approaches Flaherty to see what he needs Singer comes rushing out of the back room with the now-hot tray of food.

Singer, knowing this is a policy violation, starts to work on his defense when Thornburg understandingly cuts in and says, "Howard, it's fine. Anytime he wants something warmed up just do it. And if anyone has a problem with it tell them to see me." Flaherty nods thanks to Thornburg and Thornburg, stone-faced as always, nods back, returning to his inspections.

Nightly inspections could sometimes keep Thornburg inside the store into the early morning, but thankfully tonight he finishes up just shy of two a.m. As he locks up the side door and walks into the chilly parking lot he hears a loud shriek emanating from somewhere near the bus stop, approximately one hundred yards away.

His first thought as the commotion grows louder are that someone is being attacked, maybe even stabbed. He cautiously walks towards the sound, not wanting to be a hero but knowing he has to do *something*.

Thornburg is stunned when he sees the noise's source. It's just Richard, sitting alone with his back against his tree. His hands frantically strike the sides of his temples as his feet kick up thick clouds of dirt around him. Flaherty's guttural shrieks reach a crescendo, as if he is locked in some sort of life-or-death struggle. Then, just as quickly as the attack escalated it subsides. Thornburg is now mortified, realizing Flaherty was actually sleeping; this was all part of some horrific visceral nightmare. Thornburg, unsure if waking Flaherty up would make it worse, walks back to his car, feeling deep sympathy for the little homeless man.

28

On the Streets

"Not all those who wander are lost."
—J.R.R. Tolkien

The subjects of PTSD and homelessness were the initial topics I was hoping to get a better understanding of when I started researching Richard's life. I wanted to comprehend how Richard — or any man or woman that served our country — could end up on the streets alone, without hope.

What went wrong? What can we do to save the vets that are still out there, and how do we prevent the new crop of vets coming home to that same fate? Why did Richard forego the conventional world and choose exile on the streets? Was it an actual choice, or was he not capable of understanding the extent of his declining mental health?

In the movie *The Ten Commandments*, Prince Ramses foregoes killing his former brother Moses for treason and instead banishes him from Egypt into the harshness of the desert. On the edge of Egypt's border, Ramses points into the vast bleak desert ahead and says,

"The slave who would be king."

Handing Moses a binding pole Ramses states, "Here is your king's scepter and here is your kingdom, with the scorpion, the cobra, and the lizard for subjects."

Whether it was justified or just his perception, I knew Richard felt the same anguish regarding his banishment from the military. Richard's desert of despair would be the city streets. His loyal subjects were no longer the thousands of insurgent fighting men he helped organize and train, but the lizards and frogs also eking out their existence in the urban Miami jungle.

Police officers have to deal with people daily who suffer from all types of mental health issues, including veterans struggling with PTSD. Unfortunately, given the recent Florida budget cuts (which downsized the mental health facilities and intervention programs) we have few helpful options for these people.

More and more of the mentally ill are being pushed out of hospitals and treatment facilities back onto our streets. Inside one of this country's oldest jails — the Dade County Jail, located in the heart of Miami — I witnessed some of the most horrific living conditions I've ever seen for the mentally ill housed there.

There are no movies or television shows that can replicate the horror, danger, smells, and utter despair that occur on that fourth floor. My heart goes out to the men and women who have to work in those conditions: I don't think I could last a week.

It's filthy, hot, and reeks of every bodily fluid a human being can excrete. The dark hallways are filled with endless shrieking echoing off the metal cages, literally the sound of madness. The worst cases I've seen up there are the ones who rigidly stare through the bars with vacant eyes, completely extinguished of life.

In Florida I was governed by a mental health law called The Baker Act.

The Baker Act is a tool for law enforcement officers, doctors, and mental health practitioners to legally intervene on an individual's behalf.

To "Baker Act" an individual is a serious thing, because you're basically taking away all their rights, removing their freedom, and forcing them to enter into a mental health facility for at least a twenty-four-hour evaluation hold. After the first twenty-four hours the attending physician can decide to either release them or hold them for an even longer period of time, as they see fit.

If I was dispatched to or just happened to make contact with an individual during my shift that I believed was suffering from mental health issues, I would first need to establish if they met the criteria of The Baker Act.

The first question I would have to answer is, "Are they a danger to themselves or anyone else? Or if I leave them untreated will they eventually cause themselves harm based on their own neglect?"

To establish these criteria I would first try to get the complainant, family member, friend, or witnesses to explain their concerns to me. I would then try to build a rapport with the individual and observe them as best as I could. After building a rapport, I would ask them if they wanted to hurt themselves or anyone else. I would ask them basic cognitive questions, like what day is it? What year is it? Do you know where you are?

Keep in mind, living on the streets or acting or thinking different than the norm is not only not a crime — it's not a reason to deprive someone of their freedom. A Baker Act must be a danger to themselves or others.

Richard at no time caused me or any of the other officers I worked with to deem him a Baker Act. The only other remedy we could offer Richard was access to shelters and mental health advocates, which were quickly disappearing.

To understand Richard's and other veterans' homelessness, I needed to understand more about his PTSD and the possible treatments available. I asked the men who served with Richard, hoping to understand what their experiences were. Why did they believe they were able to hold it together while Richard eventually resorted to living on the streets? I later asked the same question of his family and friends.

Captain Rick Lencioni
101st Airborne

"I've personally been in treatment with the VA for, God, I don't know ... maybe twenty years or so? The first time I went to counseling I was in between marriages, and this friend told me, "You are borderline suicidal." He said, "You're out of control. All you're doing is putting yourself at risk over and over again." And my answer to him was, "I like walking on the edge of the razor." Because it was true at that time I was constantly putting myself at risk of being killed or seriously injured.

"My friend directed me to this guy who was a behavioral psychologist specializing in trauma, and this was before PTSD was really identified — he was also a Vietnam veteran. He got into behavioral psychology because of it, and was helping other veterans. He helped me.

"He helped me understand what was wrong. He helped me realize that I had something wrong. That was the biggest thing: I thought I was fine. I didn't think there was anything wrong with me. Then, the truth hit me square in the face. I went to see him and started crying like a baby within the first ten minutes. That woke me up. Then I dropped out for several years and went back to my old ways. When I hit rock bottom again, I went back to see him for more help.

"But it's just something that you have to recognize, and you have to make yourself do. The problem with a lot of Vietnam veterans, or veterans of any war, is they don't recognize they have the problem. So, they don't do anything. No, I'm fine — there's nothing wrong with me.

"I went through one-on-one counseling for a few years, and then they put me into a group where there were probably fifteen or twenty of us. We'd sit around and shoot the shit. Sometimes it got into war experiences, but a lot of times it was just how do you deal with something that's going on in your life? You're looking at it in this particular way, and the reason you're doing that is because of your Vietnam experiences.

"You're looking at it the wrong way. You're assigning either the wrong level of importance to it, or you're perceiving a threat that's not really there. And the other guys in the room would pick up on it because they're not using your perspective, but they can talk to you and steer you back to the right way. And that worked really well. I did that for probably seven, eight years easily.

"Look at the U.S. population. What is it, like one half of one percent that's served in the military? And how many served in actual combat, going through the violence of warfare as Richard experienced it? So, when Richard put himself out there, nobody walking up and down that street understood a thing about him. You didn't understand anything about him because he was not telling you, or anybody else for that matter. We as a generation didn't tell anybody: that's just how we were.

"There's no way to ask for help when you're not talking about it. Richard didn't have anybody to talk to, so he's out on his own. He's just living the way he thinks he's supposed to live, and the people walking around him are oblivious to it. They just think he's some homeless drunk.

"You know, I did some volunteer work for a while. I did it with my daughter over at Metropolitan Ministries here in Tampa. You get homeless people who show up in there because they need something, they need a box of food or they need a blanket. And every once in a while you'd meet someone really interesting.

"Most of the time when I went over there I would serve them popcorn and coffee and soft drinks, stuff like that. We had a little cafeteria area. Well, they'd open up in that environment. They're sitting there, they're having a cup of coffee, they're having some popcorn, and I'm standing there, and they want to talk, and they would talk to me. You'd be surprised at the stories to come out of that.

"But you know, you can't save everyone. A person has to be able to participate in their own rescue. If somebody's just dead weight, nobody's going to save them. You can't do it. So, for Richard to have been saved Richard would've had to want to be saved. If he didn't want it, it didn't matter what any of us would've done. Even if we'd done it all together we weren't going to save him, because as soon as we let go of him he'd go right back out there on the streets. It's tough to get Vietnam out of your head — you're always reliving parts of it."

"After the war you seem to have been pretty successful. What do you think kept you on the right track?" I asked.

"I can't say for sure, probably family. My father was a Chicago police officer and I had a very strong feeling that I had to impress him. I couldn't let him down, and to do that I had to be responsible. I ended up with a family. I got married for the first time while I was still on active duty, so I had a wife and I thought *Well, you can't screw this up.*

"I won't say I didn't make a lot of mistakes. I made a hell of a lot of mistakes. My mistakes were survivable, fortunately. They didn't have to be survivable, but I was lucky. Richard may not have been so lucky. Richard's mistakes may have taken him down a dark path. When I came back from Vietnam I stayed in the Army for a little while like he did. While you're still in the Army you're okay because you're surrounded by your peer group, who respect you and who think you're a hero. Because you come back and these young officers that haven't been there yet are going, 'Wow, look at all the fruit salad medals on his chest.'

"So, you're still good. You're not dealing with that civilian population out there thinking you're a piece of crap, right? So, you come out of the Army and now you're dealing with these civilians. I couldn't take dealing with them for years. It took me many years to get over the fact that I hated civilians. I thought they were undisciplined, that they had no idea where they were going in life.

"Like I said, I continued to make mistakes just like Richard — and I'm sure he made a lot of mistakes. But for some reason I kept pulling myself back up and going forward again. I just kept going forward. I never did get a college degree, never got to do that, but I ended up Vice President of a major corporation. The truth is I don't know how I got there. I just somehow managed to do it."

"Somehow you kept the train on the tracks," I said.

"Fell off a few times. You know, where we had to get the big industrial crane out to pick me up and put me back on the tracks. But it worked out in the end. And then I married my second wife, who helped me. Lori and I have been together thirty-three years, and we're just solid. I'm sure the relationship made a big difference in that I couldn't let her down. Sadly, Richard probably didn't have a relationship like that.

"I went through a lot of dark times too, but my demons might've been maybe not as bad as Richard's. He might've created a few extra for himself by his attitude. It's hard: the shock of leaving combat and coming back to the United States. All of a sudden everything's peaceful.

"Leaving the war, it's like you woke up from a bad dream — it's instantaneous. You land here after that freedom bird takes you home, you're looking around and you're going, "Wait a minute. I can't just walk across that road over there without checking the flanks. I have to take my time first to feel everything that's under my feet, for every step I take could be my last."

"But when you look around nobody else is doing that. You're the only one doing it. You live like that for a while until you learn to acclimate back into civilian life — or back into peacetime life, at least. And it's a shock. You don't really get over it for a very long time, if you get over it at all."

Sergeant Ron Kuvik
101st Airborne, "Echo" Company

"That was a shock to me when I found out Richard was just living on the streets. So, yeah, I was shocked. I would never have expected that of him, because he really seemed to have his life in order. He seemed to be always in charge. Seeing him interact with other people in the platoon, and the way he talked to me, it seemed like he had a really good idea of who he was. I would never have expected Richard Flaherty to become a homeless-type of person, and so the only thing I can think of is that he chose to be."

"How much of Richard's frustrations do you think were due to coming home to a country divided?" I asked.

"It's a part of all of our anger. When we were in 'Nam, we had an enemy over there. The North Vietnamese, and the Vietcong, that was our enemy. When we came home we felt as if we were betrayed by our government, betrayed by the media, and betrayed by the population, by our country. So, if you want to know who I thought my enemy was when I came home, I would say, 'You're my enemy.' People who stayed here, civilians who stayed home. When I came home I was subject to a lot of ridicule, verbal abuse. I'm sure Richard was too. Other guys that I served with in 'Nam came home to the same thing.

"Well, I also served under Rick Lencioni, and I was proud to serve under Rick. Rick was the same kind of guy as Richard: big heart, a man of integrity, but always a warrior. And so, it was a pleasure to serve with both men and have them as my leaders. Even though they had different leadership styles out in the field, I would've taken a bullet for either one of them. That's just how strong our bond was. I'm sure they both would have done the same for me.

"That's what we were there fighting for. It wasn't for honor and the flag so much as we were fighting for each other. Yeah, it was part of that too. There we respected and honored our country. We took an oath. We had the hearts of a patriot, but yeah, we're fighting for one another too. I don't remember any slackers over in 'Nam ... none. Everybody pulled their own weight.

"For me, the scariest thing I faced was coming home. That was scarier to me than being in 'Nam. And I was so surprised when I came back to my hometown to see nothing had changed. That surprised me so much, everything being the way it was before — that's because I had changed so much. I was scared the first time I went out in public after I got back, I shook. I didn't have my buddies around me. I didn't have my M-16. I had no grenades. I had nothing. I was defenseless, yes, and I just shook. Just scared the crap out of me. It took me a while, and I don't think I've ever really adapted to living in society again.

"There's a part of me that just likes to get out of society for a while, likes to get out of touch. I used to drive my wife crazy during the first few years after I came back from 'Nam. We had a place — a house in the country. It was woods all around us; I'd walk outside on a Saturday afternoon and without even thinking about it, just without any plan to do so, I would go for a walk in the woods and maybe along some streams, or just go and come back the next day without planning to do it.

"I'd go out and stay overnight and she'd be so upset with me when I got home. But it was just that I wanted to get out of there. There were times I wanted to walk away from everything. Of course I didn't, because I had a wife and children. I couldn't walk away from my responsibilities, but if I didn't have a wife and kids maybe I would've. Maybe I would have been sitting next to Richard by that tree.

"I do miss the jungles of 'Nam though, because there was an honesty over there. The bullets didn't lie. You lived more life in one day in 'Nam than you do in six months over here in the world. You fought with everything you had. Out there with the men, you were really alive.

"To this day I skydive, and I'm sixty-eight years old. I've got eleven jumps so far this year. I'll go out in my kayak, out where the alligators are and the sharks out on the ocean, or whatever. Just I don't know, it's something in there that makes you feel more alive. Or like that Corvette of mine, having it go a hundred-twenty or one-hundred-and-thirty miles an hour.

"When I finally came home from 'Nam I came home injured on a litter. But after I got healed up I wanted to reenlist and go back. You know, I wanted to go back and go back and go back, but they wouldn't let me reenlist because of my injuries. They retired me, and I was devastated. You want me to do what ... be a civilian? I don't know how to be a god damn civilian. I'm not a civilian! You changed me ... now I'm a savage. A warrior ... a savage in my heart ... and I still am, so don't piss me off!"

Carl Cain
10st Airborne, "Charlie" Company

"I lost a lot of friends over there, and I still have a hard time making friends with people because I'm afraid something will happen to them. A lot of guys when they came back from 'Nam — they was messed up. Still a lot of them are messed up for life. I almost became an alcoholic when I first came back, never did any drugs, but I could've gotten strung out on drugs if I let myself keep falling. I owe a lot of my savior and recovery to my family; that's what kept me strong during the bad times."

Russel "Doc" Hall
101st Airborne Medic

"The guys told me when I got back, 'You need to go to the VA and get some help.' Since then I have gone, because I got PTSD due to certain things that happened to me. I can imagine what Richard went through being an officer — you're responsible for your platoon. You have to do what you think is right. And every decision Richard made over there in the heat of the moment he now has to live with. That's a heavy weight to bear. I can't imagine the pain of losing a man that you were responsible for. I just can't imagine it."

Al Dove
101st Airborne "Echo" Platoon

"I loved that little guy, and would have done anything for him over there. It breaks my heart thinking of that proud man, a hero, living on the streets. It's a damn shame. I lost touch with him after the war, but I always wondered whatever happened to him, hoping he was having a great life.

"You know, when I went over to 'Nam I was an illiterate. I couldn't read or write, but I could read a military map and could do almost anything with my M60 machine gun. I could shoot from the hip and take out a running target a hundred yards away. My M60 kept men alive, they all counted on me back then. Here in the States I don't have that gun anymore, and it sometimes feels like no one needs me. Maybe Richard felt the same."

Walter Flaherty
Richard's Older Brother

"You start carrying a lot of scars emotionally from the war and the fighting ... those scars built up in Richard, and there eventually was a breaking point. It's the wounds that he had incurred both inside and out. Because he would get hit with bombs and get blown up out there — and I mean, you get tossed around — you're not really the same after that. You've got bruises, and deeper wounds.

"They are finally starting to talk now about the effects of the bombs and the traumatic brain injuries these men sustained. You see it now in football, how bad those concussions are. It just wasn't understood back then.

"Once we set him up in a little apartment in a retirement community, but he saw it as a handout as opposed to help and went back to the streets. He just wasn't comfortable sitting in one spot. I think he had the nightmares of the war, you know, and it would get to him from being in one place."

Donna Marlin
Richard's Cousin

"I pleaded and I pleaded with him to come to live with me. I'd say, "Why don't you just come and stay here? I love you, Richard, and I can't have you living on the streets. It's tearing me up knowing that." And he'd say, "Donna, you know I would, but I can't stand the cold." But I don't think that was it. I think he was homeless because that was his freedom."

Fred Gleffe
ATF Agent and Friend

"I think to this day the reason why Richard was always so bitter and why he became homeless was because he got RIFFed out of the United States Army. You know, the reduction in force. He just always said that he really enjoyed being in the military, and in Vietnam. He said he never wanted to leave. He said, "I would have stayed over there until they kicked me out." He always was talking about how he missed wearing a uniform every day, how he missed the camaraderie. He missed just being out there and being useful."

Dennis Connors
Childhood Friend

"Why was he homeless? It was the war. It was from losing his guys. Yeah, he told me he lost three guys in the war. You know, he told them to go this way and they did something else, and they wound up getting shot. But you know, I guess that really upset him. Richard was a senior officer. When you give an order you expect people to carry it out. It wasn't his fault, but I guess he carried that guilt with him all his life."

Chris Mancini
Federal Prosecutor

"In the military, especially in times of war, everybody's expendable and nobody's special. You can be replaced. Here's a guy who literally put his life on the line for the government in a way, and in a way for himself. I mean, Richard was a risk-taker. He got a thrill out of what he did, you know? So, it's not like it was a one-way street. But when you stick your neck out as much as a guy like that does, being a Vietnam vet and Green Beret ... I mean, and then the government just cuts you loose?! That's tough. That's just very tough."

On November 30, 2009, Richard went to the VA hospital and met with a psychiatrist for the first time. Parts of his statement were recorded in the medical notes and read as follows:

> "I have been living on the streets for the past twenty-one years. I have never seen a psychiatrist before. I used to drink two beers and some wine per day. In March of 1975, my girlfriend died in an accident. Other women I have lived with have died, and I have increased my alcohol intake to a lot more. I feel depressed and anxious. I request this treatment for my depression and my anxiety and also help to get off the streets."

To get authorization for treatment for his PTSD, Richard was asked by the VA to provide written examples of what he saw in Vietnam:

> "We came under fire. One WIA, I flanked the enemy and killed one VC with an M-26 grenade. My fire team leader was shot in the back. He died a foot from me. I could hear his internal organs collapse and watched his body shrink a bit. Sargent Meeks was behind me. Seconds later, the M60 gunner was hit in the head.

I saw the hit. When I got to him, one third of his skull was blown off exposing his brain."

His second statement:

"Two VC refused to surrender and were armed with at least one M2 carbine. An M26 grenade was ordered dropped into the spider hole, when we dug them out it was then discovered they were two VC females about mid-twenties. They were still alive but one had a fist-sized star-shaped hole in her forehead with a piece of brain matter on top of her skull. The other had blood streaming from her ears and missing part of her fingers. I was a foot and a half away, called medevac, then found a Ziploc bag with lipstick, rouge and one ounce of perfume. This humanized the enemy, cutting my effectiveness as a leader in half."

His third statement:

"On April 20, 1968 in the Thua Thien Province I assigned Sgt. Epps' squad to take point. A soldier that I knew volunteered to take the lead. A call came in shortly thereafter that he was WIA. I rushed to the position and immediately engaged the enemy. We then made two attempts to recover the wounded lead soldier and in that time he became a KIA. During the recovery attempts I had to rush the bunker line. We recovered the body of the man I knew and his blood was on the straps of my rucksack."

There were four more handwritten visceral statements from Richard recalling what haunted him throughout his life. Richard also went on record at the VA hospital saying that he believed many of his later illnesses were due to being sprayed with Agent Orange:

> "While assigned to D company, 501st, we could see the planes spraying Agent Orange on the Laotian border. Fifteen to twenty minutes later, we could feel a mist on us. Also when I was leaving Vietnam in December of '68, I was sprayed while at Camp Sally. Same scenario: three planes in formation near the Laotian border."

At the beginning of this book, I mentioned that I had a theory why Richard chose to tell me his life story when and how he did. I believe he knew his days were numbered. Not because hitmen or State Department agents were chasing him in the shadows, but because the skin cancer on top of his head was getting worse. The treatments at the VA hospital weren't working.

Richard's cousin Donna also alluded to the fact that Richard was never scared of dying in war, or in any of his later clandestine missions. His biggest fear would be to die in a hospital wasting away.

I don't think Vietnam ever lets any of these men truly escape her grasp. I believe Richard's exposure to Agent Orange accelerated his cancer; like so many other Vietnam vets it continued to take its toll on those caught under its deadly rain.

In his psychiatric evaluation the doctor concluded that Richard had 'grandiose and paranoid delusions, poor insight, and poor judgment.' But interestingly, at the same time his cognitive function was nearly perfect; his intelligence IQ was well above normal and just below genius. In a dementia examination called the Folstein test, Richard scored just two points below the maximum — indicating, no dementia.

Other findings listed under Mental Status Exam:

> Appearance and Attitude: *Patient is well groomed, alert and awake, oriented to person, place, and time, patient is cooperative with interview.*
>
> Family Circumstances: *Father died of heart attack mother died after mastectomy surgery for breast cancer.*
>
> Financial Issues: *Receives only $376.00 a month*
>
> Substance Abuse: *Patient is an alcoholic. He actively drinks 1 ½ liters of wine a day last drink was this morning. On weekends drinks Vodka.*
>
> Recreation Activities: *Watches movies for $6 dollars on senior days and eats pizza.*
>
> Sleep: *4 hours or less*
>
> Appetite: *'Not too good'*
>
> Motor Function: *Grossly intact with excellent gait*
>
> Mood/Affect: *Mood is depressed with appropriate affect to thought content.*
>
> Thought Process: *Linear and organized. Patient exhibits no signs of suicidal tendencies and advises that "I'm a combat veteran therefore I'm always paranoid."*

Physical examination list of Richard's injuries: Damage to his ears affecting his hearing (explosions), hands (arthritis), lower back (herniated discs — numerous parachute jumps), ankle (arthritis — sprained in combat), right leg (shrapnel), left shin (shrapnel), right eye damage (grenade), and second degree burns on his arms (explosion) were noted.

Richard's lower back injury was also mentioned in the book *A Walk in the Park* by Odon Bacque. Odon and Richard went through Special Forces Training School together, and in his book Odon recalls Richard taking a hard fall during a parachute jump exercise, necessitating medical attention for his back.

29

The Last Week of Richard's Life

Richard seemed to be in real good spirits that week. Maybe it was because he was finally telling someone his life story and that was therapeutic for him, but I'm really not sure. As I look back on that week it was so strange, because it seemed everything was starting to fall Richard's way: everything suddenly looked so positive.

May 2, 2015 — Saturday
Subway Sandwich Shop

We met up at our favorite lunch spot — the Subway sandwich shop in the back of Loehmann's Plaza. The owner of the shop, Amin, was working behind the counter and told Richard how honored he was to meet him. That act of respect on the owner's part seemed to make Richard a couple of inches taller that day. Richard was never big on showing emotion, but I remember as we were leaving Richard warmly thanked Amin for the great meal and tipped his cup of soda into the air, a sign of salute.

The other thing I remember about that meeting is this one line Richard told me. Then, to make sure I wouldn't forget it he made me say it several times out loud. "Dave, the key to life and the key to all your problems is this: money in the bank calms nerves!"

May 4, 2015 — Monday
Boston Market Restaurant

We met for lunch that day at the Boston Market in the northern part of the city. As I slid into my seat the stabbing pain of my sciatica lit up my hip and ran down my right leg like a Christmas tree. Talk about pain in the ass. That morning I received my second round of epidural shots in my spine, but it seemed they were only making the pain worse.

I had my notebook with me, and after we finished eating Richard answered more questions about his life. As we were leaving something strange happened in the parking lot. Just as we said goodbye Richard looked at me very accusingly and said, "By the way, don't think I don't know what this documentary thing is really all about. I know you're working for those State Department guys, spying on me."

I was stunned, really thrown off and shaken up. I felt bad because all this time I thought we were friends, and here he accuses me of being one of those people out to harm and torment him.

Even though his accusation was highly delusional, it still sucked. If he really believed I was just another person trying to harm him then maybe I'd over-stepped my boundaries. Maybe asking a man in his condition to participate in this sort of project was wrong on my part.

"Richard, what are you talking about? You're my friend. I just want to get your story out to the world, like we agreed on."

"Okay, we'll see," he answered as he turned on his heel and walked away.

May 5, 2018 — Tuesday
Mobil Gas Station

The day before at the Boston Market we made plans to meet and have our coffee at the Mobil Gas Station Café. I sat there nervously waiting on him; after yesterday's shocking accusation I was not only worried that he wouldn't show — I was worried he would never talk to me again.

Richard walked into the café area right on time and sat right down on a stool next to me. He was his usual friendly self and wanted to get right back to work. He acted like that State Department conversation never happened, so I went along and didn't bring it up either.

After hearing what Frank Sosa said in his interview with my dad, it all makes more sense. When he talked about Richard always testing people, always pushing buttons just to see how they'd react ... that's got to be what his accusation was all about. Apparently I passed the test sufficiently enough for Richard to continue to trust me. As we moved ahead with the project nothing was ever spoken about that subject again.

May 6, 2015 — Wednesday
Zone 2

While I was working my shift I got a call from Richard, asking me to meet him at an apartment building in the neighborhood. He wouldn't explain what it was about, just told me that if I had time to swing by — he would be there for maybe another half hour or so. We patrolled neighborhoods in zones, and that building was in my zone. As long as I didn't get a call for service I would be able to meet him.

The apartment was on one of the higher floors, and

the front door was open. It was a modest unfurnished one-bedroom condo facing onto a golf course. As I walked inside I saw Richard talking to a woman dressed in business attire. He then introduced her to me as his real estate agent. *Real estate agent — what the hell is he up to now?*

Richard motioned me to the kitchen and I followed. He shook his head disappointedly as we looked at the older appliances and counters. "Definitely going to need to be renovated," he said.

"Rich, what are we doing here?"

"They're asking one eighty-five, but I think I can get them down to one sixty. It's been on the market for a while."

Now I was really confused. Richard never mentioned a thing about buying an apartment in all our countless conversations — where in the world would he get the money? Richard must have seen the skeptical look on my face. He added, "My old friend Abe Saada owes me some favors. He offered to help me buy a place."

Okay, let's carefully go over this scenario. In my later interview with Abe I asked him about this incident. He smiled at the thought of it and said, "Buy him an apartment? I own apartments, and you can't imagine how many times over the years I offered to rent him one at almost no cost. I'm not talking about a million-dollar condo type of place, but a nice clean little apartment in a good neighborhood was my offer. So no, I never offered to buy him an apartment."

Now the question of this puzzle falls back on me and my understanding of this enigma of a man. My thoughts: despite all his PTSD and paranoia, I don't think for one minute Richard Flaherty would randomly waste some real estate agent's time by showing him apartments if he didn't have the money. It just wasn't in his character.

But Abe was very sincere — and why would he lie? He had no idea I was going to ask him that question, and his answer was quick and genuine. As a former detective who has done dozens of interviews and interrogations — my bullshit meter is very sensitive. I didn't get a hint of deception from Abe.

My only conclusion was that either someone else (who Richard didn't want me to know about) was buying him the apartment, or he had the money stashed away in some offshore account that I earlier speculated about. Saying Abe Saada offered to buy him the apartment was just Richard's cover story in case people asked how a home-less man could suddenly afford to purchase a condo.

Perhaps at almost seventy years old the streets were getting too tough on him. Maybe he was finally able to access some squirreled-away money. I don't think I'll ever know the real answer to that mystery, but I'll file it as just another incredible sadness that he lost his life when he was so close to getting off the street.

May 7, 2015 — Thursday
Richard's Palm Tree

I was close to the end of my shift and I wanted to swing by Richard's tree to check on him. I found him in his usual posture with his big backpack wedged up against the tree, him reclining on it with his arms folded. I walked up to him with my phone video already recording, but saw he was deeply asleep. I can clearly remember the peaceful look upon his face, which I'd never seen before. We've all seen adults in that deep state of sleep, reminding us of the innocence of a child. In that cocoon the weight of the world is lifted; in dreams, anything is possible.

Deep sleep wasn't something I remembered ever seeing Richard enjoy, so I made sure to back away quietly. As I was about to leave a small voice told me to take a picture of him. I barely had any actual photographs of him, and for some unknown reason we never posed for a picture together. I walked to the other side of his tree and took a single photo.

It wasn't until months later that I looked at the picture again, noticing something incredible. The street lamp positioned behind Richard appeared in a white glowing beam of light; it pointed to where Richard's body would be lying only one night later.

After twenty years of police work and constantly being around people who have already passed (or being with them at the exact moment they pass away), I can honestly say I've never seen any proof of the afterlife or ghosts, or anything that pertains to the supernatural. That image of the rays of light inside the picture I took of Richard is just something I can't explain.

May 8, 2015 — Friday
Aventura Police Station

Earlier that week before our afternoon roll call session started I told some of the guys at work about me doing a documentary on Richard's life. One of the officers who took a real interest in Richard's story was Officer Nelson Reyes. After completing twenty years at an adjacent police department in the North Miami Beach P.D., Nelson transferred over to Aventura for a fresh start. Being new to the area Nelson didn't know anything about Flaherty, and he was beyond touched when he heard about a war hero living on our city's streets.

After roll call and before hitting the streets Nelson questioned me on what we could do to get Flaherty into a shelter, or whatever else we could assist him with. I brought him up to speed regarding Flaherty's reluctance to go to a shelter and his general aversion to people getting involved in his personal affairs.

I told Nelson the best thing we could do for Flaherty

was help him with his ongoing fight with the VA hospital. I explained how he'd been petitioning for years for medical benefits, but the paperwork was mired in red tape. Nelson told me he would make some phone calls and do what he could. *Good luck,* I cynically and regretfully thought as I walked away from him.

A couple of hours later Nelson called me with the following news. He'd just spoken to a friend, a woman named Habsi Caba. Habsi was head of the Miami-Dade crisis intervention team for the eleventh judicial circuit. Richard's story also touched Habsi, and she wanted to do everything in her power to get him immediate help — even if she had to call the governor (who she personally knew and had a good working relationship with) to get the ball rolling. *Damn, things are finally starting to go Richard's way!*

It was great news. I knew Richard would be ecstatic. Finally, allies on his side to help him fight what he felt was the injustice of the VA. hospital! I excitedly asked Nelson, "When can we introduce Richard to Habsi? What's the quickest we can get this off the ground? Do you think next week would be too soon?"

Nelson, in his commanding tone of voice, stated, "Dave, you don't get it. Habsi isn't going to wait till next week — she wants to meet Richard as soon as possible. Can we meet up with him later tonight so you can introduce me to him?"

"Of course, I'll call him right now."

Although Richard was homeless he did actually have a cell phone. It was one of the older flip phones, but it worked. He used it sparingly.

I called Richard. He didn't pick up, so I left him a voice mail. With time to kill I decided to call retired ATF agent Fred Gleffe for the first time and check on Richard's story about working undercover in that federal case. I then called Gleffe; our conversation is the first event I write about in this book.

I finally got a call back from Richard around 8:30 p.m. He told me that he'd just left Aventura and was in Sunny Isles Beach. Sunny Isles is a small beach town located just a half-mile to the east of Aventura. When I asked him what he was doing, he told me he was sipping some cocktails on the beach watching the waves roll in. I told him I wanted to meet up with him because I had good news. He said he was going to jump on a bus and should be back at his tree at around 9:30 p.m.

At 9:30 I went to his tree and he wasn't there. This was probably the first time Richard didn't show up as scheduled, but I didn't dwell on it because I could always speak with him tomorrow. I called a disappointed Nelson and told him we could try to meet up with Richard tomorrow, since he wasn't around. Little did I know that within the next few hours he would be dead.

11:50 P.M.
Miami-Dade Police Headquarters

A thin woman in her mid-fifties, dressed in casual attire, works her way through the secured darkened hallways of the Miami-Dade Police Department. She exits the building and walks into the familiar parking lot. She gets into her silver Toyota Prius and drives it out of the parking lot, heading to her home.

12:33 A.M.
Palm Tree

Flaherty, sleeping under his palm tree, is woken up by the faint sound of a helicopter. As he rubs his eyes and stands up he looks into the star-filled sky. Seeing the small dark outline of a helicopter moving away from him, he shoulders his backpack and starts to cross the empty street at the crosswalk. He follows in the direction of the barely visible helicopter.

A pair of headlights emerges from the blackness as the Prius turns east-bound down the same street towards Flaherty. Just as Flaherty is about to reach the median, he sees the Prius' headlights — but it's too late.

A loud explosion of shattering glass and twisting metal mixes with the heavy thud of a one-hundred-pound man being struck by a car traveling twenty-five miles per hour.

12:53 A.M.
N.E. 199th St. & 29 Ave

The now-empty and quiet city street is interrupted by the sound of high heels brusquely striking the pavement. Papers and debris from Flaherty's backpack blow across the road like tumbleweeds.

The woman driver of the Prius quickly walks west-bound on the sidewalk of the street, just twenty yards north of where Flaherty's body lies. The woman looks into the street — towards Flaherty — in an attempt to locate something she is searching for. Despite the fact she passes within twenty yards of Flaherty's twisted and bloody upper body she doesn't stop or approach him, but keeps walking.

Later that morning a crowd of curious onlookers watches the police and detectives working inside the yellow taped-off crime scene. The detectives finish up covering Flaherty's body with a yellow blanket just as the first news vans arrive on the scene. Mixed with the crowd of inquisitive onlookers the driver of the Prius stands quietly by, observing the officers.

In *The Giant Killer* documentary film I thoroughly covered all the events, evidence, and legalities of the hit and run case. If it was an option I would choose to leave this section out of the book, because it now seems pointless; however, this book is not for me. It's for the readers. As such, it would be unfair of me not to try and answer at least some of your questions.

One thing I tried as best as I could in the documentary was to be unbiased. I didn't think the documentary should be used as a platform for me to express my politics and opinions to the world. This project was always about honoring Richard Flaherty, and gleaning what lessons I could learn from his life's cautionary tale.

Despite the bizarre timing of Richard being killed only hours after I called agent Gleffe, I don't think my phone call had anything to do with his death.

As a former detective, I always believed that extraordinary accusations require extraordinary proof. To this day I have not found any evidence leading me to conclude that the driver of the Prius intentionally struck and killed Richard, that this was all some type of planned hit.

If you're going to explore the idea of it being an assassination, let's think about the incredible amount of logistics needed to accomplish the task. How many people would it take to coordinate this perfectly timed hit?

You'd first need to have someone watching Richard twenty-four, seven. You would then have to coordinate with the "hitman," perfectly timing their driving to match the exact sequence of Richard getting up and crossing the street. Hardest of all you would have to pull this off in a business district, hoping that no witnesses would inconveniently walk or drive by and quickly call the cops.

Now, theoretically, you could set up the timing so it wouldn't be by chance. You could have someone call or implore Richard to cross the street at the exact moment the driver entered the area. That's theoretically possible but tough to implement, because Richard was always so cautious.

But then, how can you explain why Richard didn't see the car coming? You would think the only way it could've surprised him was if he initially saw the car and realized it was far enough away that he was safe to cross — but then the driver sped up, or switched lanes at the last moment when Richard wasn't looking.

What about the idea that Richard was intoxicated, that he was the one at fault? The medical examiner did find a significant amount of alcohol in Richard's blood. However, as a functioning alcoholic it probably wouldn't have affected him too much. Even in the middle of the night he had enough good judgment to cross the street at the crosswalk before being struck.

Maybe the driver had her lights off? The evidence shows she most likely did have her lights on. Only a hundred yards later she crossed through a camera light, recording her car right after the crash. The recording shows her right headlight is on while her left headlight is off, probably due to it being broken during the collision.

How could she claim to have immediately walked back to the scene and not seen Richard's body? Sure, it was dark and the lower half of his body was concealed in the bushes, but his upper body was clearly visible — explain that? And didn't she know exactly where the impact occurred on the street, so she would know exactly where to look? Also, why return to the scene in the morning, look at the yellow crime scene tape and the yellow blanket covering Richard's body, but not tell someone what happened? There were more than enough police officers and detectives standing right there for her to talk to. Lastly, what about the lack of tire or skid marks?

My best guess on the entire incident would be she initially had no idea that she hit him. This was due to her being in some type of trance-like state caused by either extreme fatigue, illness, or some type of medications or substances, legal or illegal. For example, maybe she ingested large amounts of Ambien or something similar that night and was operating solely on autopilot.

Anyone that has ever hit a deer or a trashcan knows it's a loud, traumatic impact on your vehicle. At approximately twenty-five miles an hour in something as small as a Toyota Prius, you can imagine how powerful and violent the impact with Richard must have been. Even hitting a large bug on the highway makes a loud sound. I'm positive her whole car buckled and shook as she plowed into Richard's body that night. Yet she never swerved, braked, or pulled off to the side of the road. Instead, she just kept on driving.

It is also hard for me to explain why she didn't see the blood and hair mixed with the extensive damage to her car. The car had damage to the front left fender and left headlight. The metal driver's side A-frame — which is the bar that holds the windshield — was bent inward, and the small driver's side window where Richard's head impacted was smashed.

The surveillance video from her parking lot shows her parking her car and then looking at the front-end damage under the parking lot lights. Why didn't she see the blood and hair? Unless her eyesight was severely compromised, I don't really have an answer to that one.

The surveillance video also appears to show that even with high heels on she wasn't swaying or showing the usual signs of impairment as she walked. That doesn't completely rule out her being intoxicated, but she certainly wasn't showing any of the clear-cut signs of being heavily impaired.

Later that night, at approximately 2:30 a.m. she calls her insurance company and reports the damage to her car. She initially states that she thinks she hit a palm frond, then changes her mind and says it was probably something else because of all the damage it caused. A palm frond in Florida is basically a discarded husk that grows on top of the palm trees and eventually falls off. It can be from a foot to several feet long, weighing anywhere from a pound to thirty pounds.

She tries several times in that recorded phone call to get her insurance company to send someone out the next day and have her car fixed. She makes no effort after calling her insurance and reporting the accident to then call the police department and report the accident.

We only know that she decided to turn herself in later in the day, after she arrived at her job at the Miami-Dade Police Department's Homicide Unit. Yes, I forgot to mention: she actually works as a court stenographer for their Homicide Unit.

For the record, despite Harvey Arango and Sergeant Jeff Burns' insistence on arresting the driver for the hit and run, the Dade County State Attorney's office refused to press any charges. It further cautioned my department that it would be opening itself up to the civil liability of a false arrest if an arrest were made without their blessing. And there you have it: case closed.

30

Riddles in the Warehouse

"Lunatic Fringe... I know you're out there..."
—Red Rider

June 10, 2016

Even after a month of going through Richard's storage locker I was still only scratching the surface of all the documentation. Some of the more curious items I found were numerous documents revealing that Richard, in the last seven years of his life, was wiring money to a woman in Thailand. Not huge sums of money, but between fifty to one hundred dollars every few months or so.

I hired a local Thai investigative service to help me track her down. A neighbor advised the investigator that the woman they were looking for was in her late fifties; she'd left Thailand around 2016 to move to Germany. The trail goes cold from there, and there's no further information on her.

I also found a Thank You card addressed to Richard. When I opened the colorful card it simply read:

Dear Mr. Flaherty,

I can't thank you enough for your random act of kindness.

It makes me believe in the goodness of humanity. Wishing

you a wonderful holiday season!

Cheers,
Nicole Graham

I called Ms. Graham. She explained to me that she lost
a bag (which included her wallet) while out shopping.
About a week later she received a package in the mail
with her bag and a note from Richard explaining where
he found it. All her ID and credit cards were still inside.

I also found a similar letter of thanks from a man who
lost his wallet on the beach. Once again Richard sent it
back fully intact, with a nice note.

———————

Everything I thought I knew about Richard was turned
upside-down by the contents of a stained, medium-sized
cardboard box I found buried under a musty bag of blan-
kets.

Inside the cardboard box was basically everything
needed for a professional spy-kit: a mini-cassette recorder,
maps from around the world, two brand new unopened
cell phones still in their boxes, language translation books
(Spanish-Arabic), small writing pads filled with handwritten cryptic notes, travel agency receipts, and travel itiner-
aries — and finally Flaherty's passport.

I checked the passport first. It was stamped full of exotic
locations from around the world: Iraq, Cambodia, Jordan,
Venezuela. And all the traveling dates were from the last
twenty years. *Holy shit!*

That's when I remembered what Richard's friend Frank
Sosa told my dad about Richard working missions.

*"Richard called it the juice, the high from walking on
the razor's edge. Richard told me when he was working*

missions he felt young again ... especially when he was out of the country. All his problems melted away and his nightmares would stop."

All this time I'd known Richard, not suspecting a thing, and he was *still* running missions. But Sosa also said Richard had many ways of covertly getting out of the country so it wouldn't appear on his passport. How many more of these missions (or 'trips') did he actually go on?

I went home with this new box of mysteries and compared it to the documentation I already had separated into files labeled as miscellaneous and travels. The following is a dateline I compiled via the various documentation:

Flaherty's Trips and Significant Documentation

2003: Grand Forks, North Dakota – Hotel

2004: Birmingham, Alabama

2005: Tok, Alaska & Yukon, Canada (Feb–April)

2006: Seattle, Washington. Sends FOIA letter to Justice Department referencing ATF documents

2006: Osceola County, Florida – Bonnet Creek Resort (near Orlando)

2007: Sends Fax to Embassy of Republic of Iraq

2007: Files an Army DA-160 form requesting immediate deployment to Iraq as a Kuwaiti Adviser – Special Forces Liaison Officer

2008: Grand Forks, North Dakota – Hotel

2008: Purchases a plane ticket for Amman, Jordan

2009: Puerto la Cruz, Venezuela – Passport Stamp

2009: Republic of Iraq – Passport Stamp

2010: Amman, Jordan – Passport Stamp

2010: Thailand – Passport Stamp

2010: Cambodia – Passport Stamp

2010: Singapore – Passport Stamp

2012: Venezuela – Passport Stamp

2013: Boston, Massachusetts – Plane Ticket

Receipts revealed that from February 28, 2003 to March 6, 2003, Richard traveled to Grand Forks, North Dakota and stayed at a small local hotel. I contacted the hotel (which had already changed to a new name); luckily the manager was a long-time employee of both businesses. She found records on the computer revealing that five years later, in 2008, Richard stayed at the hotel on the same exact days as in 2003.

What was so important in North Dakota? Why would Richard — who was known for hating the cold — head up there in the dead of winter, when the temperature averaged twelve degrees Fahrenheit? I called some of the local reporters in North Dakota to see if they could dig up any significance to those dates. I was thinking that maybe there was a big gun show or some other military event going on, but when they got back to me they informed me nothing of any significance could be found on those dates.

While working my off-duty job at the movie theater I happened to meet a retired Army Ranger (Kris "Tanto" Paronto) who had experience working all over the world with private military contractors. Many of his missions were to provide security for the CIA and U.S. ambassadors. I met Kris at the red-carpet premiere of the movie *13 Hours* directed by Michael Bay. Usually when film director Michael Bay came to our theater for a red-carpet event I would be assigned to him as his security. Kris was one of the men working in Benghazi during the terrorist attack on September 11, 2012, as a CIA private military security contractor.

As a former Ranger, Kris was very familiar with Richard's world. Knowing from his own experience how bad North Dakota gets in the winter, Kris speculated that Richard could have been in the area visiting one of many military bases. "There are a lot of missile silos up there, and back in the cold war days that's where all the nukes were."

I also spoke to a retired intelligence officer, who speculated that Richard might have been crossing the border unnoticed up into Canada through North Dakota. He remembered a time when various "agencies" kept safehouses in the area as jumping off points for missions.

In 2004 Richard called his Vietnam vet friend Carl Cain, who was living in Alabama. He told Carl that he was close by in Birmingham and would try to stop over. Richard didn't mention what he was doing in Alabama, only that he would call back to arrange a time to visit. Richard never called back; that was the last time Carl ever spoke to Richard.

In 2005, Richard headed up to Alaska and stayed in the Yukon, Canada, and Tok, Alaska areas. On the surface the trip would seem to be for pleasure, but once again why go in the middle of February and stay until April?

In 2007, Richard sent an official Army DA-160 (request to reenlist) form for immediate deployment into Iraq. Following the capture of Saddam Hussein Iraq was experiencing a combination of power vacuum and the mismanagement of the occupation, leading to a lengthy insurgency against U.S. forces. In 2007 the United States responded with a troop surge.

Insurgent warfare was Richard's specialty. As a staunch patriot, I'm sure he felt it was his duty to help train young

American soldiers in the lessons and tactics he learned thirty-eight years earlier. I've never located any documentation or testimony leading me to believe that Richard was ever officially reinstated into the Army.

In 2008 Richard purchased a plane ticket for Amman, Jordan. This leads me to believe he was planning to use Amman to bypass sovereignty and cross into Iraq. The paper trail stops with his plane ticket. There is no stamp in his passport revealing if he made the trip or not.

At the end of 2008 Richard writes a letter to his Vietnam buddy, Carl Cain, and advises him that he will soon be traveling to Venezuela. It appears Richard was very concerned about this trip; in the letter he asks Carl to serve as his emergency contact "If things go bad in Venezuela."

In 2009 his passport revealed that he entered Caracas, Venezuela, and then traveled to Puerto La Cruz. Puerto la Cruz is a small port city known for its lawlessness and weapons smuggling. Experts of the region advised me that Richard would need a liaison in-country because it would be too dangerous for him to enter and travel alone.

Did Richard renew his weapons sales relationship with the Venezuelan generals he did business with in the late seventies when working for Bushmaster Rifles? Newspaper records reveal that on November 29, 2006, Venezuela received their final shipment of one hundred thousand AK-103 assault rifles, purchased from Russia a year earlier. The rifle deal was part of a multi-million-dollar military build-up by the oil-rich country. The sales became a source of heated rhetoric, mostly from Chavez, and strained relations with the U.S. Unhappy with the Venezuelan government's lack of cooperation on terrorism and concerned about its military build-up, the Bush administration in May of that year banned U.S. weapons sales to Venezuela, pressuring numerous other countries to likewise refuse to export weapons.

Was Richard in Venezuela as a black-market arms dealer providing weapons to the Chavez government, or was he sent to Venezuela for some type of intelligence gathering mission? If you look at Richard's history of being a devoted patriot, along with his hatred of any form of communism or dictatorship, I think it makes the possibility of him knowingly providing the Chavez government with weapons highly unlikely. Until I go to Puerto la Cruz myself, or till someone from the government comes forward, I doubt I'll ever solve that mystery.

Also in 2009, Richard's passport was stamped as he entered Iraq through the Consul, Hikmet Bamarni. There is no further documentation or witness testimony of this excursion. The most likely answer once again was provided by Kris Paronto. Kris advised me that many private security companies purposely hire older Special Forces soldiers to train the local security or police forces. Not to do any of the actual field work, but to serve as teachers and advisers only.

The advantage for the companies in hiring older guys — besides their experience — was that they tended to blend in more naturally with the local populations. Being low profile would raise fewer eyebrows with the locals than a younger, in-shape and recently retired Special Forces soldier.

The last interesting thing Kris unknowingly told me about military contract work was that the State Department ultimately runs most of it. His exact quote was, "The State Department has its fingerprints on damn nearly everything."

He added that the State Department would technically be the agency responsible for tracking American citizens when they travel abroad. They'd certainly have access to all of Richard's passport information.

When I asked Kris his opinion on Richard's paranoia about State Department men following him around, Kris was surprised. Kris was friends with many older Special Forces soldiers like himself, who suffer from what he called PTS (not PTSD, or Post-Traumatic Stress Disorder).

Kris didn't categorize it as a disorder — he felt it should be known as just Post-Traumatic Stress. Kris was familiar with the paranoia a lot of these men experienced, but he never heard any of them accuse the State Department of spying on them. It was always the CIA or FBI.

In dealing with his own post-traumatic stress, Kris said, "I was diagnosed with it in 2007, but I know how to deal with it. I continued to deploy with it overseas for six years and went through Benghazi with it. I now do these speaking events where I have to relive a particularly bad event, and it really drains me. But you know what? Going back overseas is where you feel normal, and that's where he may have found some peace.

"I'm sure Richard went back because when you go overseas and you work, the post-traumatic stress isn't there anymore. You're back in an element where your adrenaline's always way up, whereas in the United States it's right in the middle. And we don't like to live like that — I'm sure he was the same way. I'm sure because of his past work, because of what he did, that he went over there to be normal, to feel like he did when he was in Vietnam, and I get that."

In early 2010 Richard's passport revealed he flew from New York's JFK airport to Queen Alia Airport, Amman Jordan. Once again I would have to assume he was there to covertly cross into Iraq.

In mid-2010 his passport revealed he traveled to Bangkok, Thailand. From March of that year until July Richard stayed in South-East Asia, traveling from Thailand to Cambodia and Singapore.

In 2012 Richard returned to Venezuela. Interestingly, in that same year Venezuela passed a new gun law banning the commercial sale of firearms and ammunition.

Until then, any citizen with a gun permit could buy arms from a private company. Under the new law only the army, police, and certain groups like security companies would be able to purchase arms. If we go back to the most likely assumption (that Richard was selling arms), the next question would have to be where did all the money go?

In Richard's storage unit he kept his bank statements from throughout the years. Although there were times when he had a few thousand dollars saved, there were never any major fluctuations. Of course, we could then speculate if Richard made huge profits selling arms he would have hidden the money in some off-shore accounts. Unless an international banker comes forward with that information, I don't think I will ever discover any hidden treasures.

Finally, in 2013 he purchased a ticket to Boston for unknown reasons. That ends the trail of his documented traveling (although again with Richard's clandestine traveling habits, this might just be the tip of the iceberg).

"He never was specific about his trips," his cousin Donna Marlin remembered. "He just said he went on his trips because people owed him money, and that he had a lot of enemies. I thought to myself many times, *Richard, what could you have possibly done to have all those enemies? I mean you're such a great guy, what is this all about?*"

His cousin Jeanie Rinaldi remembered about his traveling, "He told me words to the effect that, 'I'm going to Cambodia to take care of some of my enemies.' I didn't press him for answers because I really didn't want to know anything more about this subject. I told him Richard, don't tell me anymore."

Aventura Police Sergeant Robert Myers also spoke to Richard throughout the years when he would see him on the streets. He told me of a particularly interesting encounter:

"One day I saw Richard walking around the streets carrying this briefcase. You know, a black leather attaché-type of briefcase that a businessman on Wall Street would carry. So, I stopped Richard and asked him what he was doing. He said that he had to leave the country, that he was going to Iraq, working for the CIA. Now I thought that was one hell of a joke, so I just told him, "Okay Richard, good luck with that," and I drove away."

In Richard's storage locker I located several receipts from his trips. Handwritten on the back are what appear to be account numbers and requests for reimbursement. Who was he requesting payment from?

I wrote two letters, under the Freedom of Information Act, to the State Department and CIA requesting information about Richard Flaherty and his travels. Both letters were returned with the same statement: "We can neither confirm nor deny any knowledge of Richard J. Flaherty."

The last item from his cardboard spy-box was the mini-cassette recorder, which looked to be in good condition. There were several unopened tapes next to it in the box, but there was one tape that looked used.

I popped the used tape into the cassette player. Maybe this was finally it. Maybe this was the big breakthrough I was looking for, where Richard would reveal all his life's mysteries and answer all of my questions. Maybe I'd even find out who really shot Kennedy, or where Jimmy Hoffa was buried.

I hit the play button. All I heard was a mechanical sound, similar to a weed cutting or leaf blowing machine used by landscapers. This annoying sound at wavering levels continued for the entire length of the tape, which was about forty-five minutes. *Nothing — absolutely nothing.*

I flipped over the cassette and started listening to side B. Wouldn't you know it? More landscaping sounds. I pushed the fast forward button — you could still hear the track in sped-up version.

Finally, there was a break in the noise. I released the

fast forward button and let it play at regular speed. I focused as hard as I could, trying to pick up any hint of Richard's voice. Then, low and behold, I heard Richard speaking clearly, as if he were standing right next to me. What he said will baffle me for all eternity. He clearly says the following:

"This is going to be my last entry for now because they're starting another episode of Matlock."

That was it. There was nothing else on the tape.

31

A Hero's Transformation

Approximately three weeks after Richard's death the Medical Examiner's Office filed their final report. They wouldn't release his body until their investigation was complete. Aventura police Sergeant Burns called Richard's brother Walter about Richard's body being released. Walter had already been in contact with the Army, and was advised Richard was qualified for the highest military burial that could be bestowed upon him. He would be receiving a military procession with a full platoon escort into the prestigious Arlington National Cemetery.

Every military man I spoke to said that Arlington was the ultimate honor. They all wanted to eventually make it there for their final rest.

Certainly, for a man like Richard — who dedicated his whole life to the military and proving that he belonged — this would be his final affirmation. However, it wasn't the ending Richard ultimately chose.

Letters that I found (sent to Richard by the Army) proved Richard already knew five years before his death that he was qualified to be buried at Arlington. However, two years before passing Richard purchased his own burial plot in a small secluded cemetery in West Virginia.

Inside an envelope where Richard kept his paperwork were receipts from the cemetery and a map showing that Richard's final wish was to be buried next to his old love, Lisa Davis. There was also a letter he wrote to Lisa's sister, explaining his feelings for Lisa. In the last paragraph of the letter he wrote:

> I've been in love with Lisa for thirty-three years.
> I will be in love with her for the rest of my life
> on this plane and beyond for death will never
> separate my love for her.

Richard J. Flaherty

The mythological Hero's Journey popularized by Joseph Campbell and Carl Jung states that the hero — after enduring trials and tribulations from a lifetime of adventure and hardship — will eventually return home. But the hero is no longer the same. Flaherty's final choice proved he no longer needed the accolades he yearned for his whole life. He finally found peace within himself, reaching his apotheosis.

It could also be interpreted as this fiery little man's last act of defiance. After living a life on his own terms, he would be the one to ultimately choose his eternal resting place. And maybe the military and government had let him down one too many times.

32

Deeper into the Jungle

*When I die and they lay me to rest, Gonna
go to the place that's the best. When I lay me
down to die, Goin' up to the spirit in the sky....*
—*Norman Greenbaum*

I'll always remember the last time I met with traffic homicide investigator Harvey Arango at the street where Richard was killed. It was approximately four weeks after Richard's death, and we spoke about the medical examiner's final report. The report revealed that Richard wasn't killed immediately upon impact, but was alive for several minutes in a semi-conscious state.

I walked over to the median and knelt in the dirt where my friend's body had lain.

Images of those grisly crime scene photos burned in my memory. He was alone in the darkness, lying on his back on the blood-soaked earth, with his right arm outstretched and his right hand open. Before losing consciousness he must have been reaching upward for something in the sky. But what was he reaching for? I would like to think in those last few moments Richard James Flaherty finally found some measure of peace.

There is no sound, only a bright flash of light. Flaherty's body tumbles through the air, landing almost fifteen feet away from the point of impact. His first thought is, *I must have stepped on a Vietcong land mine.*

His bloody and broken body is lying awkwardly in the dirt. He attempts to sit up, but his body is not responding. He tries desperately to reach for his rifle, knowing the enemy will soon be coming to finish him off. His blurred vision turns the nighttime jungle foliage into hues of green and black.

For a moment he allows himself to slide down into the dark abyss, finally accepting that it's time to let go — but a voice implores him that his men still need him, and he claws his way back up to the surface.

He forces his eyes to open and commands his body to sit up, but he can't: his weight is already too sluggish, too heavy to move. He fights with his last ounce of strength to slightly crane his head to the side to get some bearing of where he's lying.

In a moment of clarity his vision clears. The jungles of Vietnam turn into the familiar streets of Aventura, Florida. The strain becomes too much, and he permits his head to return to a comfortable position, allowing him to gaze up into the night's sky.

He once again hears the faint sound of that Vietnam-era Huey helicopter.

The familiar Vietnam jungle smells of rotting vegetation and pungent foliage start to flood his senses.

On the empty dark Aventura Street a vintage Huey helicopter descends towards Flaherty.

Twenty-three-year-old Airborne Sergeant Ron Kuvik, in full jungle fatigues, stands at the open helicopter door pointing down towards Flaherty.

The helicopter hovers over the median a few feet above Flaherty. The propeller wash flattens the bushes around him, causing newspapers and street debris to go flying.

Kuvik kneels and extends his arm out of the helicopter,

reaching for Flaherty. He yells over the prop noise, "Grab my hand."

Flaherty, with his last remaining strength, reaches his arm up towards Kuvik. As soon as he grabs Kuvik's hand Flaherty transforms into the once-again youthful twenty-three-year-old Green Beret Captain.

Flaherty is pulled into the helicopter and sees all his old comrades. With a big relieved smile on his face he takes one last glance out the door.

"Captain Flaherty, it's time to go home."

The helicopter lifts off and heads into the darkness.

> *"Vietnam vets don't die.... We just go deeper into the jungle."* —Ron Kuvik

Captain Richard J. Flaherty was awarded the following medals:

Silver Star
2 Bronze Stars with Valor
2 Purple Hearts
Parachute Badge
The Gallantry Cross
Combat Infantry Badge
2 Army Commendation Medals with Valor
Vietnam Service Medal
Vietnam Campaign Medal
The National Defense Service Medal

On December 18, 2012, a fifteen-foot bronze statue
of Richard's cousin Medal of Honor recipient
Homer Wise was erected in Stamford,
Connecticut's Veteran Park.

Located inside the Fairchild Botanical Gardens in
Coral Gables, Florida, is the Lisa Davis Anness
Butterfly Park, open daily to the public.

The Florida State Attorney's office never charged
the government employee who killed
Richard J. Flaherty and left the scene.

On June 6, 2015, Captain Richard J. Flaherty
was finally laid to rest in a small secluded
cemetery in Milton, West Virginia,
next to the woman he loved.

In mid-2018, after almost twenty years of police work,
David Yuzuk retired due to his back injury.
Later that year he fulfilled his promise
to his friend Richard and released the documentary,
The Giant Killer, to the world.

Dedicated to the memories of

Captain Rick Lencioni

Captain Walt Yost

Abe Saada

Sergeant Ron Kuvik

General John H. Cushman

Trinity Catholic High School
Richard J. Flaherty

RICHARD FLAHERTY

The giant killer . . . clever and witty . . . newspaper . . . college bound.

64

From the Trinity Catholic High School yearbook

1/501 Airborne "Screaming Eagles"

Officer Candidate School (OCS)

Cu Chi Vietnam: 2nd Lt. Richard Flaherty, Al Dove, and Jerry Austin

Richard's father and his brother Walter Jr. "Timmy" at Pellicci's Restaurant

SECTION I

THE SILVER STAR MEDAL

1. TC 320. The following AWARDS are announced.

FLAHERTY, RICHARD J 05347431 (SSAN: 042-36-2344) SECOND LIEUTENANT INFANTRY
Company C 1st Battalion (Airborne) 501st Infantry APO San Francisco 96383
Awarded: The Silver Star Medal
Date action: 20 April 1968
Theater: Republic of Vietnam
Reason: For gallantry in action in the Republic of Vietnam on 20 April 1968.
Second Lieutenant Flaherty distinguished himself while serving as a
platoon leader of Company C 1st Battalion (Airborne) 501st Infantry.
Company C was involved in combat operations in the Quang Dien District,
Thua Thien Province, Republic of Vietnam. At 1140 hours, Second Lieutenant Flaherty's platoon, the lead platoon, was taken under intensive
automatic weapons and RPG fire. Second Lieutenant Flaherty, realizing
the seriousness of the contact, immediately maneuvered his platoon to
deliver a flanking assault against the enemy positions. Throughout
the battle he repeatedly exposed himself to the hostile fire in order
to better direct the suppressive fires of his squads. Lieutenant
Flaherty immediately called a 90mm recoilless rifle team to his position after having spotted an enemy bunker position to his front
which was delivering automatic weapons fire on his platoon. Lieutenant
Flaherty then personally directed and assisted the 90mm recoilless rifle team in an assault of the enemy bunker, braving the intense hail of
hostile fire. Under Lieutenant Flaherty's astute direction and leadership, the enemy bunker was swiftly destroyed, enabling his platoon to
advance and continue its devastating attack against the enemy. Second
Lieutenant Flaherty's extraordinary heroism while engaged in close combat with a well dug-in enemy force was in keeping with the highest traditions of the military service and reflects great credit upon himself,
his unit, and the United States Army.
Authority: By direction of the President of the United States under the pro-
visions of the Act of Congress established 9 July 1918.

Silver Star — April 20, 1968

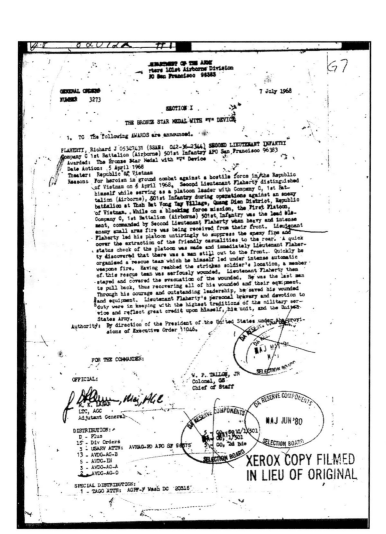

DEPARTMENT OF THE ARMY
Hqrs 101st Airborne Division
PO San Francisco 96383

G7

GENERAL ORDERS   7 July 1968
NUMBER   3273

SECTION I

THE BRONZE STAR MEDAL WITH "V" DEVICE

1. TC The following AWARDS are announced.

FLAHERTY, Richard J 05347431 (SSAN: 042-36-2344) SECOND LIEUTENANT INFANTRY
Company C 1st Battalion (Airborne) 501st Infantry APO San Francisco 96383
Awarded: The Bronze Star Medal with "V" Device
Date Action: 5 April 1968
Theater: Republic of Vietnam
Reason: For heroism in ground combat against a hostile force in the Republic
of Vietnam on 6 April 1968. Second Lieutenant Flaherty distinguished
himself while serving as a platoon leader with Company C, 1st Battalion (Airborne), 501st Infantry during operations against an enemy
battalion at Thah Hat Vong Tay Village, Quang Dien District, Republic
of Vietnam. While on a blocking force mission, the First Platoon,
Company C, 1st Battalion (Airborne) 501st Infantry was the lead element, commanded by Second Lieutenant Flaherty when heavy and intense
enemy small arms fire was being received from their front. Lieutenant
Flaherty led his platoon untiringly to suppress the enemy fire and
cover the extraction of the friendly casualties to the rear. A quick
status check of the platoon was made and immediately Lieutenant Flaherty discovered that there was a man still out to the front. Quickly he
organized a rescue team which he himself led under intense automatic
weapons fire. Having reached the stricken soldier's location, a member
of this rescue team was seriously wounded. Lieutenant Flaherty then
stayed and covered the evacuation of the wounded. He was the last man
to pull back, thus recovering all of his wounded and their equipment.
Through his courage and outstanding leadership, he saved his wounded
and equipment. Lieutenant Flaherty's personal bravery and devotion to
duty were in keeping with the highest traditions of the military service and reflect great credit upon himself, his unit, and the United
States Army.
Authority: By direction of the President of the United States under the provi-
sions of Executive Order 11046.

FOR THE COMMANDER:

OFFICIAL:

_____, Maj, AGC   W. P. TALLON, JR
                                               Colonel, GS
M. K. LESAR   Chief of Staff
LTC, AGC
Adjutant General

DISTRIBUTION:
  D - Plus
 15 - Div Orders
  3 - USARV ATTN: AVHAG-PO APO SF 96375
 13 - AVDG-AG-B
  5 - AVDG-IN
  3 - AVDG-AG-A
  2 - AVDG-AG-O

SPECIAL DISTRIBUTION:
  1 - TAGO ATTN: AGPF-F Wash DC 20315

XEROX COPY FILMED
IN LIEU OF ORIGINAL

Richard sleeping under his palm tree
in Aventura, Florida

On the night before Richard's death street light
points to where he would be killed

First time meeting Captain Rick Lencioni
in Tampa, Florida

First time meeting Walt Yost
in the mountains of North Carolina

Roger Mathis   David Yuzuk   Walt Yost

Meeting Sgt. Ron Kuvik

Interviewing 101st Airborne's Carl Cain

Interviewing Army Ranger Kris "Tanto" Paronto

Interviewing ATF Special Agent Fred Gleffe

Gleffe in the 80's during the Cocaine Cowboy
days of Miami

For the reenactment scenes in the documentary I needed to find both a young and an older version of Richard Flaherty. I found professional jockey Arnyy Salgado at the racetrack and Oscar Montoya working maintenance at the Aventura mall

Richard's footlocker and medals

Donna Marlin showing me Richard's uniform

Neil Yuzuk interviewing Vietnam vets at the 101st
Airborne reunion in Nashville, Tennessee

My dad Neil Yuzuk giving me a pep talk at Long Hunter
State Park, Tennessee

Richard's childhood friend Rick Farina showing
David Trinity H.S. yearbooks

David and Neil presenting the Richard J. Flaherty
plaque to Ernie Boucier for the Trinity Catholic
H.S. Wall of Heroes

At the statue honoring Richard's cousin Homer L. Wise
recipient of the Medal of Honor

U.S.DEPT.of the ARMY   Feb. 28, 2005
AWARDS and DECORATIONS

Re: RECOMMENDATION M.O.H.

SIR or MS:

AROUND MID-APRIL 1968 IN THE QUANG-TRI PROVINCE OF THE REPUBLIC
OF VIETNAM, THE 1ST BN.501ST INF.,2ND BDE, 101ST ABN. DIV. WAS ATTACK-
ED BY NVA UNITS.(M.I. EST. TROOP STRENGTH TO BE ABOUT TWO THOU-
SAND) SIMULTANIOUSLY, ALL 4 LINE COMPAMIES, THEIR AMBUSHES, THE
BN. H Q./FIRE SPT. BASE,AND THE RECON PLT. WERE ATTACKED.

'C'Co. HAD 2 AMBUSHES, ONE SQUAD & ONE PLT. SIZE. THE SQUAD SIZE
AMBUSH SPRANG EARLY EVENING. THE PLT. AMBUSH WAS OVERRUNNED.
PLT. SGT. KIA, PLT. LEADER SERIOUSLY WOUNDED AND OVER 20 WIA'S.

LT RICHARD JAMES FLAHERTY, PLT LEADER, 3RD PLATOON, ACTIONS
THAT EVENING WERE COURAGIOUS, UNCONTESTABLE AND ABOVE & BE-
YOND THE CALL OF DUTY.

OUR SQUAD AMBUSH(3RD PLT.)SPRUNG FIRST. WE WERE POSITIONED TO
BLOCK THE MOST LIKELY AVENUE OF ADVANCE. S.O.P. FOR AMBUSHES ,
LT.FLAHERTY'S, WAS TO FIRE ALMOST THE BASIC LOAD. THIS SHOW OF
FORCE CONVINCED THE NVA TO ABANDONED PLAN 'A' AND ATTACK
FROM ANOTHER VENUE. A FATAL MISTAKE FOR THEM.

IT WAS ABOUT 1:00 A.M. WHEN LT. FLAHERTY ORDERED FULL ALERT FOR
THE PLT. SHORTLY THEREAFTER 3 R.P.G.'S SLAMMED INTO THE C.P.TEM-
PORARLLY KNOCKING OUT COMMO. SHRAPNELL FROM THE 3RD RPG. HIT
LT. FLAHERTY IN THE HEAD LODGING IN THE TEMPLE AREA.DESPITE
BEING WOUNDED WHEN A FLARE WENT OFF IN FRONT OF THE 2ND PLT.
AREA AND NOONE RESPONDED, LT. FLAHERTY FIRED FOUR ROUNDS IN
THE DIRECTION OF THE LIGHT. SECONDS LATER ALL HELL BROKE LOOSE.
OVER TWO HUNDRED EXPLOSIVES WERE THROWN AT THE PERIMETER;
BUT THE NVA FAILED.TO BREAK THROUGH. THE 1ST PLT. AMBUSH WAS
NOT SO LUCKY. THEIR CLAYMORES AND MOST OF THEIR TRIP FLARES
WERE CUT, EXPLOSIVES THROWN DIRECTLY ON THEM, AND THE NVA
OVERRAN THEIR POSITION RAKING THEM WITH AK-47 FIRE. THE ONLY
CASUALTIES AT THE COMPANY LEVEL WAS 3RD PLT. MACHINEGUNNER
KIA AND ONE WIA FROM THE AMBUSH.

One of Richard's men recommended him for
the Medal of Honor

LT. FLAHERTY CALMED EVERYONE DOWN WHILE HE ORGANIZED A DE-
FENSIVE COUNTER MOVE. ANTICIPATING A FLANKING MANUEVER HE OR-
DERED A SQUAD TO USE THEIR AIMING STAKES AND FIRE FROM ALTER-
NATING POSITIONS EVERY 15 SECONDS.(in the dark you would loose the site
picture off the muzzle-flash in 10 seconds) THIS ACTION DEPRIVED THE ENEMY
OF MOBILITY, INFLICKED CASUALTIES AND HALTED THE ADVANCE. FOR-
CING THE NVA TO RETREAT TO THEIR ORIGINAL ATTACK POINT.

LT. FLAHERTY THEN GATHERED GRENADES AND WENT TO THE PERIME-
TER AND INSTRUCKED A CERTAIN PATTERN TO BE THROWN AT CERTAIN
INTERVALS. BOTH ACTIONS RESULTED IN HEAVY ENEMY CASUALTIES AS
EVIDENCED BY SCREAMS HEARD AFTER THE EXPLOSIONSAND MASSIVE
POOLS OF BLOOD FOUND THE NEXT DAY.

WHILE THE COMPANY FARED WELL, ONE KIA, AND ONE WIA BOTH FROM
THE 3RD PLT, THE 1ST PLT AMBUSH SUFFERED OVER 65% CASUALITIES. AND
WITH THE PLT. SGT DEAD AND THE LT. SERIOUSLY WOUNDED, AN E-5
WAS IN-CHARGE. NOT KNOWING THEIR EXACT LOCATION NO ARTILERTY
OR FLARES COULD BE FIRED. AND BECAUSE OF THE EXTENT OF THE
ATTACK ON THE BN,MED-EVAC'S WOULD NOT BE CALLED UNTIL DAY-
LIGHT.

THE CPT CALLED ON LT. FLAHERTY TO LEAD A RESCUE OF THE STRICKEN
PLT. HE ORGANIZED A 10 MEN TEAM AND PROCEEDED ON AN AZMITH AS
INSTRUCKED. HOWEVER , HE STOPPED 100 METERS OUT AND RADIOED
THE PLT.TO FLASH ONCE WITH A RED FILTER ON A CERTAIN BACK AZ-
MITH. LOCATING THE PLT HE ORGANIZED A DEFENCE. AND AFTER GET-
TING RPTS ON WHAT HAPPENE, HE DETERMINED TO GET THE WOUNDED
BACK TO THE COMPANY PERIMETER.

LT. FLAHERTY, THE RTO, AND MYSELF REMAINED AT THE SITE TO PRO-
TECT WEAPONS, AMMO & EQUIPMENT. WHILE SGT GRAVES LED EVERY-
ONE ELSE BACK TO THE CO. HOWEVER, WHEN HE RETURNED THERE WAS
ONLY HIMSELF AND TWO OTHERS. LT FLAHERTY ORDERED THE RTO AND
MYSELF TO ASSIST SGT GRAVES'S TEAM AND CARRY THE WPNS AND AS
MUCH AMMO WE COULD CARRY BACK TO OUR PLT.

LT FLAHERTY REMAINED ALONE TO SECURE THE REMAING AMMO AND
EQUIPMENT. AND TO ACT AS LP/OP AND FO, IN DEFENCE OF THE
COMPANY. HE ALSO INSTRUCTED US NOT TO RETURN UNTIL AFTER THE
MED-VACS WERE CALLED IN THE MORNING. COMPANY STRENTH
AMOUNTED TO 74, 33 IN THE 3RD PLT .

LT. FLAHERTY'S ACTIONS, STRATEGIES, AND COMMAND PRESENCE
SAVED LIVES AND INFLICTED HEAVY CASUALITIES ON THE ENEMYAND
SHOULD BE AWARDED THE CONGRESAL MEDAL of HONOR.

MOST OF US IN "C" CO, 1ST BN.,2ND BDE.,101ST ABN. DIC., THOUGHT THIS
WAS ALREADY DONE. IT WAS A SURPRISE TO FIND OUT RECENTLY THAT
WAS NOT THE CASE.

AT A TIME WHEN DESERTERS AND COWARDS ARE REWARDED WITH BOOK
CONTRACTS AND TOURS ON THE LECTURE CIRCUIT; HEROS SHOULD BE
HONORED BY A GRATEFUL GOVERNMENT REGARDLESS OF THE POPULA –
RITY OF THE WAR.

Making of the Documentary

When the idea came to me to try and make Richard's story into a documentary the first person I contacted for my "team" was my dad, Neil Yuzuk. He was also the only person I could afford at that time. My dad is a retired substance abuse counselor who worked for the NYC Board of Education for twenty-two years.

My small documentary team was then joined by local film maker Jeremy McDermott. I approached Jeremy earlier on with the idea and told him about all of the perks; it was going to be a lot of hard work, and I had almost no money to pay him. Why he took pity on me and joined the team I'll never know, but I was grateful to have him on board.

Omen of Things to Come

After Richard was killed I tried to get someone from Richard's family to fly down to Miami and take possession of all the items inside his storage unit. No one from his family could make it. Instead, they asked me if I could inventory the items and contact them if anything of value was found.

Problematically, the storage unit was rented by Richard on a monthly basis. If the warehouse fee wasn't paid, within two weeks all his items would be auctioned off or thrown in the trash. Since Richard didn't have anything of value to auction, I knew it would all be tossed in the garbage ... not unlike his life. I therefore started and continue to this day to pay the monthly rental cost.

As I persisted in my research, I got tired of driving to the unit and scouring through files in the warehouse's dim light and musty air. I grabbed three of the most important boxes of documents and took them to my home to read in comfort. I placed the boxes downstairs on an end table beside the only chair I could sit in without too much pain—my electric zero-gravity recliner.

The next day I went to work. Halfway through my shift I received an urgent phone call from my girlfriend screaming for me to get home right away. When I got to my house my girlfriend was standing outside soaking wet, clutching my terrified chihuahua Pedro. I looked into my house: all I saw was a waterfall that was once my front door. My girlfriend told me she stopped by the house to check on Pedro. When she opened the front door a wave of water tumbled out up to her knees. Slogging inside she found Pedro in the living room doing his best impression of Michael Phelps, swimming hard to make it to safety atop the high ground of the couch.

Once she got Pedro safely outside she went back into the house. Despite the risk of being electrocuted she ran upstairs and luckily found the broken bathroom pipe. She shut off the main valve and stopped any further flooding.

Oddly, as I waded through the indoor pool which used to be my living room I noticed that the water from the upstairs used only one A/C vent to drain down to my first floor. That A/C vent was located directly above Richard's three boxes, which would have remained dry if they were placed only a few feet away. For forty years Richard kept those documents pristine, and I managed to ruin them in all of one day. As soon as I got off the phone with a water damage clean-up company my girlfriend and I spent the next several hours laying out hundreds of Richard's documents by my pool and on top of the bushes to dry in the sun.

For a moment I thought, That's it, I can't take this shit anymore. It's time to quit this crazy documentary; it's just not meant to be. This must be a sign. But then the strange idea of Richard purposely doing this to test my resolve occurred to me. He was just messing with me, in that way he often did with people. Grinning I said, "Screw that— I'm not quitting."

A Trip Down Alligator Alley

In many conversations during those last two weeks of Richard's life an ugly topic involving his brother Walter came up—and Richard was certainly adamant about it. Richard blamed many of his later-in-life financial problems on Walter. Richard talked about inheritances that were kept from him and all sorts of other money mismanagement. Richard told me that they hadn't spoken in over ten years.

Despite everything Richard told me (and the fact he would never have approved) the first person I knew I had to speak to was Walter. I loaded up my van and drove west from Miami across the Everglades on the aptly named highway "Alligator Alley." I turned north when I hit the west coast and headed up toward Tampa. As I crossed the picturesque four-mile-long Sunshine Skyway bridge a vast dark storm system sat out in the gulf, like a warning sign. As I drove onward brilliant orange glowing rays of sunlight shone out from behind the clouds. Was this a metaphor for Richard's life?

As I pulled into the small fishing town of Punta Gorda (where Walter lived) I started feeling some trepidation. How would this go? I had absolutely no idea what Walter would soon tell me, or even how I would feel toward

Walter once I met him. Was I being disloyal to Richard by seeking out his brother? I didn't think so. This journey was always about the truth, no matter how uncomfortable it might get.

I met Walter outside his apartment complex. He was sitting on a bench, which immediately caused flashbacks to Richard and me talking on his bench. Walter was a medium-sized soft-spoken man in his early seventies, who bore a slight resemblance to Richard. We decided to take a walk by a small pond; I noticed Walter was suffering from some physical afflictions. His body was stiff, and he had some slight tremors in his arms. Although his speech was slow and careful, his mind seemed sharp and resilient. The overwhelming feeling I got from being in Walter's presence was one of deep sadness.

In our hours of conversation I didn't detect even the slightest bit of animosity on Walter's part when it came to his feelings toward Richard. Over and over he talked about how proud he was of his little brother, the war hero. I never had the heart to ask Walter why he thought Richard mistrusted him. Walter had just buried his little brother: there was no point in me making matters worse.

Maybe the truth was Walter did hold up some of Richard's money because Walter felt Richard wasn't of sound mind and wouldn't be capable of making good financial decisions. Of course it's always possible Walter was just was as Richard said. That's a subject I refuse to make a final conclusion on, because usually the truth lies somewhere in the middle.

Meeting Captain Rick Lencioni

Listening to Richard's recollections of the war helped me understand what he saw and experienced, but to complete the picture as thoroughly as I could I relied on the men that served with him. The man who brought me into that world was the same man Richard mentioned many times while telling his stories, Captain Rick Lencioni. The cover of this book is one-half Richard's image, while the other half is Rick Lencioni. I tracked Rick down with the help of other vets and contacted him by phone. We made plans for me to drive up to his home in Tampa and meet in person.

If Hollywood were looking to cast the perfect man to exude every image and trait of the gritty war hero, they would choose Rick. I was shocked the first time I met him; as he opened his front door and I shook his power-ful hand I thought he could easily pass for a man in his early fifties. He was lean and muscular, with a full head of dark hair and a full-sized Burt Reynolds mustache. Judg-ing by his looks, it wasn't surprising when he told me that at seventy years old he was still routinely skydiving and spending long hours traveling around the country on his Harley-Davidson motorcycle.

Rick welcomed me into his home. One of the first things he told me was that he was the proud son of a Chicago Police Officer. Throughout that afternoon Rick provided me with incredible insight not only into Flaherty's world, but his own experiences in Vietnam.

Rick told me he enlisted in the Army in 1965 and went through Officer's Candidate School and Special Forces School, thereby earning his Green Beret. In January of 1968 he was sent to Vietnam as a First Lieutenant, getting assigned a platoon. Later in 1968 (after being wounded) he was promoted to captain.

His awards included The Combat Infantry Badge, Senior Parachutist Badge, Bronze Star Medal with V (v for valor) Device, Purple Heart, Air Medal, Army Com-mendation Medal, the Presidential Unit Citation, and the Vietnamese Cross of Gallantry.

Lencioni and Flaherty were alike in so many ways, yet in others polar opposites. The two men's words and expe-riences of Vietnam were eerily similar. It was, however, their life choices after the war that differentiated them: initially running parallel, then turning in opposite directions at a moment's notice. So similar to their parallel mission that day in the Eight Klick Ville, which would become a metaphor for their lives.

Rick Lencioni was the man I believe Richard Flaherty could have become if he'd made a few different choices and received a break in life's unpredictable wheel of fortune. Many of the words, stories, and feelings Flaherty expressed to me that night at the Kosher Kingdom were retold in the same words, cadence, and sentiment a year later by Rick. Destiny would ultimately lock their souls in eternity, as their images so easily blended for the cover of this book.

In 1986, Oliver Stone released his film Platoon, which explored the duality of man. In the movie, two non commissioned officers (Barnes and Elias) are locked in a battle of wills on how to best win the war. Barnes is the image of hard-nosed decisions and brutal win-at-all-cost tactics, while Elias operates by putting his integrity and morality first, even if it comes in conflict with completing the objectives of a mission. After Elias is betrayed by Barnes he still welcomes him like a brother, staying optimistic despite the horrors of war.

———————

Eighteen years earlier dueling Lieutenants Richard Flaherty and Rick Lencioni had already acted out these scenes in the jungles of Vietnam, constantly testing each other's wills and philosophies. But their goals were always the same: to win the war and bring their men home safe.

A couple of months after meeting Rick he invited me up to a reunion of the 101st Airborne in Nashville, Tennessee. I called my dad and asked him to meet Jeremy McDermott and me up there.

We arrived a day before the event started, so I figured: let's record some extra footage for the documentary of my dad and me walking through the woods. We decided to go to Log Hunter State Park—it was absolutely breathtaking. A vast blue lake surrounded by a forest teeming with wildlife, and just your occasional hiker.

As I mentioned earlier, this journey wasn't always downhill or smooth. There were many times I doubted if I could finish the project. Well, just as my gas tank was running low my dad helped pick me up. We were taking a long hike, looking for that perfect spot to shoot some footage, when we plopped down on this big log in the woods to take a short break.

"You know, we're taking a big financial risk flying out here hoping some of these guys will talk to us," I said with trepidation.

My dad nodded along, staying quiet, being the perfect listener.

"And a lot of people are telling me I'm crazy for doing this, that I'm just wasting my time and money. They ask, 'Who do you think you are? What's a cop doing making a documentary?'"

"You should have just said, 'I'm a man of many talents,'" my dad advised.

"Yeah, I guess. Wasn't the guy who created Star Trek ... what was his name?"

"Gene Roddenberry," my dad shot back.

"Right. Gene Roddenberry—wasn't he a cop?"

"Yeah, he was," my dad said with a smile. "Dave, don't keep looking at the finish line and worrying about if you'll ever cross it. Just try to just keep your goals small for now—one step at a time—and things will start to come together on their own.

"Let's just focus on this weekend, get a couple of good interviews done and see if we can learn anything new. We›re on the right path. It's going to happen!"

The next day at the event before entering the large hotel convention room, I was still nervous. Although we were invited as Rick Lencioni's guests, I was unsure if any of the men would talk to me. I was a stranger trespassing in their world, so young that I was born the same year they were fighting in the Tet Offensive.

My dad had made up some flyers, which included a picture of Richard and some of his basic info. We passed them out on the first day of the event. The men were cautious but polite, sort of eyeballing us at first to see what we were all about. By the second day we had dozens of vets coming over to us and welcoming us. They not only told us about their encounters with Richard, but also shared some of their own harrowing personal experiences.

One man who made a big impression on me was Sergeant Ron "The Koov" Kuvik. Ron served under Richard when Richard ran Echo Company's Reconnaissance Unit, better known as Recon. I met him that first night at the bar, and we both put down our share of Jack and Cokes toasting warmly to the memory of Richard.

The next day I interviewed Ron. He gave me a brutally raw and powerfully honest take of his experiences in Vietnam, as well as coming home to a country that turned its back on him. With his steely arctic blue eyes and scowling grin, he let loose some of his most primal thoughts. His respect couldn't shine any brighter for his lieutenant and friend Flaherty, and his disgust couldn't be any deeper for those who called him a baby killer. Articulate and charismatic, Ron Kuvik would be the perfect voice to represent the 'I ain't going to take no more shit' attitude of vets of his era.

Like many of the other vets I met on my journey, I kept in touch with Ron. He was even warming up to the idea of letting me take his prized Corvette out for a spin. Sadly, on

April 5, 2018, Ron passed due to a sudden heart attack. Rest In Peace Sgt. Kuvik, and soft landings.

Later I would travel to San Antonio Texas to interview Carl Cain, Richard "Tex" Laraway, and Russel "Doc" Hall, who all served under Richard in the 101st Airborne. My dad would head back to California and speak to Jerry Austin and Al Dove, who served under Richard in his Echo Company Recon Unit.

When I tried to find out about what happened to Colonel John H. Cushman, Richard's commanding officer in Vietnam who was eventually promoted to General, one of the vets at the reunion gave me an over ten-year-old phone number for him in Washington D.C.

I called the phone number, hoping I could at least contact a relative or someone who knew him—if still alive the general would be close to ninety-five years old. This gruff-sounding voice answered the phone, and when I asked if he was a relative of General Cushman or if he could direct me to someone who knew of him the man on the other end of the phone loudly barked, "Relative? I am General John H. Cushman!"

Well, we spoke for a while. That's where I learned he called Richard his one-meter lieutenant.

General Cushman was still sharp as a tack. He even challenged me with, "What's a cop doing making a documentary?" We had a couple of laughs, but he didn't have too many independent recollections of Richard beside saying he was a fine officer and that his little one-meter lieutenant really took it to the enemy. Sadly, six months later on November 8, 2017, General John H. Cushman passed.

We lost Richard Flaherty on May 9th 2015 and almost exactly three years to that day on May 10th 2018, Captain Rick Lencioni lost his battle with cancer. This book would never have happened without Rick's help. This book is dedicated to Rick Lencioni and his family Lori, Rickki, and Jameson.

After Walt invited me up to North Carolina I flew to Knoxville, Tennessee, rented a car, and started driving east toward the mountains. As I drove through the winding scenic roads of the Great Smoky Mountains, it occurred to me that it had been years since I actually took a vacation. That's the thing about police work: it consumes you with a fog, years flying by in the blink of an eye. If it wasn't for Richard putting me on this path I might have procrastinated for even more years before seeing this incredible part of our country.

This project, which was supposed to be about me trying to unravel the mystery of Richard's life, was becoming more of an awakening for me. In the last ten years—in between buying a house, a divorce, and constant back problems—I'd been missing out on too many opportunities for self-growth. I had become a creature of habit, that ant following the same trail, and now all of a sudden I was doing something I hadn't done in years—being spontaneous.

I spent the next two days driving through small towns in the mountains, often stopping to talk with the locals. On my way through Bryson City I decided to do some exploring and randomly drove down a few of the more rural mountainside roads.

Farms and beautiful cabins appeared every several miles or so, between incredible forests and fast-moving rivers. Just as I was about to turn my car around a man walked onto the road in front of me, waving at me with a hand gesture that implied he wanted me to drive up to him.

I don't believe in ghosts, but at fifty yards away and with the sun in my face I swore I was looking at Richard. Although he was wearing glasses and his gray beard was a little longer, he was the spitting image of the man. I drove my car slowly forward, believing this hallucination would soon disappear. Instead the man—with Richard's same disapproving look—impatiently waved at me to hurry up.

I pulled up to the man. Because of his height he easily looked straight into my passenger window, waiting for me to roll it down. I rolled the window down and he gruffly said, "I need a ride into town." Without waiting for my answer he pulled on the door handle. I quickly unlocked the door, thinking, Well, let's see what happens now.

He hopped in. I realized this wasn't a ghost—but damn did he look and sound like Richard. I tried to make conversation, asking, "Car break down?" With a slight southern accent he answered, "No. Haven't had a car in years, although I'm fixin' to buy one."

For the next ten miles I told him the reason I was in the mountains, and a little about Richard.

"You're saying he was a little fella like me?"

"Yes, sir. May I ask how tall you are?"

"I'm five feet tall, give or take an inch."

"What type of work do you do?"

"I've done many jobs, but I mostly work as a plumber. My size actually helps me get into spots normal fellas can't get into."

Before I could get in my next question we'd arrived in town. He held up a hand and said, "Right here will be fine."

I will never forget what he told me as he got out of the car. "By the way, my name is Gene. I was also a soldier in Vietnam. Next time you're in town you can interview me, 'cause I also got some stories." My jaw dropped, and he walked away with his back to me in the same exact manner Richard would walk.

In 2016 I flew north to a hotel in an undisclosed location to finally meet Special Agent Fred Gleffe. Even before meeting him in person I had a tremendous amount of respect for Fred due to the types of cases he worked throughout his career. Fred worked as an agent in Miami during the wild cocaine cowboy days, even arresting one of their most dangerous hitmen, Jorge "Rivi" Ayala. Fred would later tell me that for a stone-cold killer "Rivi" was actually very polite and easy to talk to.

He also arrested other interesting characters, like world-renowned mercenary David Thompkins. Thompkins was offered a contract of ten million dollars by the Colombian cartels to kill Pablo Escobar while Escobar was sitting in jail. Thompkins planned on purchasing an A-37 Dragonfly light attack jet aircraft, which he would use to blitz the prison with several five-hundred-pound bombs.

Fred, working undercover, helped set up the deal with Thompkins to purchase the jet, along with several bombs. After Thompkins paid a twenty-five-thousand-dollar deposit to undercover agents he was tipped off about the sting operation and fled back to his home in Britain. He was finally arrested in Texas at George Bush International Airport after trying to sneak back into the country. From the wild streets of Miami to squaring off around the world against Bulgarian hitmen, Fred was another once-in-a-lifetime person I encountered while on this journey.

I met Fred in the lobby of my hotel—and Fred was a pretty big-sized guy. He was close to six foot four, with a large muscular frame to go with that height. My immediate thought was how odd Fred and Richard must have looked when they worked together for all those months.

We headed up to my hotel room, where Jeremy had already set up all of the recording gear. Fred brought several large folders filled with paperwork that he immediately started organizing as soon as he sat down.

You can learn a lot of things when you first meet someone by their mannerisms. It was immediately apparent that Fred was highly organized, meticulous, and took immense pride in his work. Here we were thirty-four years later, and Fred still kept all his paperwork and notes in pristine condition.

For the next several hours Fred methodically went through every detail of the undercover case. At the end of the interview I asked him again if he would be willing to talk about any of the classified information on the case.

"Dave, I'm sorry but I'd rather not speculate or comment on anything classified that Richard might have told you about the case," Fred quickly answered, and that was that.

Mental Health

Regardless of all the incredible feats Richard accomplished, there was one underlying thread that ran through my entire investigation: the issue of mental health.

I believe it's a topic Americans need to have a more serious conversation on. We now obviously know how important early intervention is for our returning vets, but we also need to look at the rise of the mental health crisis in our civilian population. Estimates now predict that more than 8 million Americans are suffering from serious psychological distress, or SPD. This crisis causes immense emotional and economic stress on both family members and the community.

"Spondylolisthesis," my doctor quietly announced. I'd been waiting the last ten days for the results of an MRI of my lower back. The doctor had a grim look on his face as he spoke the diagnosis.

"If it were just the ruptured discs I wouldn't hesitate to go in there and clean them up, but with your Spondy you're going to need a full fusion of several of the vertebrae."

The "Spondy" meant that one of my lumbar vertebrae had slipped forward. This slippage also caused a crack, or stress fracture, in another portion of my vertebrae called a spondylolysis. The fusion procedure he discussed was essentially a welding process, fusing together my affected vertebrae so they would heal into a single, solid bone.

At close to my twentieth year in police work, this was a big decision I would need to make. It wouldn't be my first surgery rodeo: I'd already had five shoulder operations. Two I attributed to my years of playing high school and college football, and the other three resulted from injuries received on the job as a police officer. My department was very accommodating for the last few months by placing me on a desk as a background investigator, but I knew I would need to give them an answer soon on my status.

My journey into researching Richard's life was coming to an end alongside my police career. It had been close to three years now since Richard was killed, and things were changing. They tore down the Publix Supermarket and the old library Richard frequented and loved, replacing them with fancy new ones. The median where Richard was killed had also been changed several times. A plaque the local firefighters made for Richard and placed on a sign close to where he died was quickly removed.

The only thing left is the tree he called home, which still stands and shades whoever is lucky enough to lie under it.

At least several times a week I pass the area where Richard lived and died. I unapologetically took on the quest to tell Richard's life story as best I could. During my journey I believe I've had more failures than successes, but I kept pushing on.

I can still hear Richard constantly quoting Winston Churchill in his repetitive fashion to get his point to stick: "Success, my boy, is going from failure to failure without losing your enthusiasm."

I made a promise to the man that I would finish what I started, and I do feel some level of closure. However, there will always be a part of me arguing that I could have done more, I could have done it better, and for that my soul won't rest. This journey—that took me all over the country—opened my eyes even more to the tragedy of our nation turning its back on its returning soldiers. The pain I saw in those men's eyes close to fifty years later will continue to haunt me. Hopefully, history will never repeat itself in that incredible injustice. I'm very grateful for all the new friendships I've made along the way and the lessons I've learned.

Now I have to ask myself: where do I go from here? To be honest I'm really not sure, though Richard gave me some direction with his endless preaching of living a life of no regret. He'd always challenge me to think about my legacy and ask, "What do you want to leave the world when it's your time to go?"

———————

In conclusion I would like to thank three last people for their assistance in getting my book published. My friend Harvey Arango, who helped me find my voice as a writer; My project manager Heather Shaw and her incredible staff at Mission Point Press; and finally my editor Scott J. Couturier.

In Scott's final summation of editing my book he wrote,

"Ultimately it seems the truth is there IS no truth to be had: just endless tunnels running off into the dark, subjective experiences, hints and whispers. This is key to *The Giant Killer's* power – it doesn't try to answer what can't be answered, offering false sentiment to cover up for loose ends."

Scott's term "endless tunnels" sparked an image in my mind summing up the essence of Richard J. Flaherty. Vietnam was Richard J. Flaherty and Richard J. Flaherty was Vietnam: a historic land of unique beauty, mystery, and danger. On the surface Richard was unique, mysterious, and certainly dangerous. However, underneath it all lay a dark maze of endless tunnels that may never be explored or understood.

Ironically, history teaches us that in January of 1966 the 25th Infantry Division constructed a base camp in Cu Chi on an Old Alluvium terrace with intent to locate – and destroy – VC tunnel activity in the area. However, the camp was accidentally built over a huge network of tunnels the VC subsequently used to wreak havoc on their above-ground overlords. Such is life and war.

Resources for Veterans

Please note that we are only listing various resources for veterans and/or their families. We do not endorse any resource listed. They are listed in no particular order.

WEBSITES AND HELPLINES

MAKE THE CONNECTION
U.S. Department of Veterans Affairs
https://maketheconnection.net

MAKE THE CONNECTION – SUICIDE
U.S. Department of Veterans Affairs – 800.273.8255
https://maketheconnection.net/conditions/suicide
www.mentalhealth.va.gov/suicide_prevention/

NATIONAL VETERANS FOUNDATION
Vet to Vet Assistance – 888.777.4443
https://nvf.org/veteran-resources/

DAV – DISABLED AMERICAN VETERANS
www.dav.org/veterans/resources/

HOMELESS VETERANS
U.S. Department of Veterans Affairs
www.va.gov/homeless/resources.asp

THE SOLDIERS PROJECT
www.thesoldiersproject.org/resources-for-veterans-and-their-loved-ones/

VFU – VETERANS FAMILIES UNITED
https://veteransfamiliesunited.org/ptsd-resources/

REAL WARRIORS – FAMILY SUPPORT
www.realwarriors.net/family/support/preventsuicide

VA SERVICES FOR HOMELESS AND AT-RISK VETERANS
www.va.gov/homeless/housing.asp – 877.424.3838

NATIONAL ALLIANCE ON MENTAL ILLNESS – VETERANS
www.nami.org/veterans – 800.950.6264

ALCOHOLICS ANONYMOUS
www.aa.org/assets/en_US/p-50_AAandtheArmedServices.pdf
www.aa.org – 212.870.3400

AL-ANON – FOR FAMILIES & FRIENDS OF ALCOHOLICS
https://al-anon.org/

NAR-ANON – FOR FAMILIES & FRIENDS OF SUBSTANCE ABUSERS
www.nar-anon.org/

VETERANS ALCOHOL & DRUG DEPENDENCE REHABILITATION
www.benefits.gov/benefit/307 – 877.222.8387

Coaching Into Care
888.823.7458 (Mon.-Fri. 8:00a – 8:00p – EST)

Crisis (Veterans) Hotline
800.273.8255 – Press 1

Gambling Hotline
800.522.4700

GI Rights Hotline
877.447.4487

Homeless Veterans Hotline
877.424.3838

In Transition
800.424.7877 (Personal Coach)

Reachout Hotline
800.522.9054

Tobacco Hotline
800.784.8669

Military One Source
800.342.9647

VA Assistance for Medical Benefits
877.222.8387 (Mon.-Fri. 8:00a – 8:00p – EST)

VA Caregivers Support Line
855.260.3274 (M-F 8a-11p; Sat 10:30a-6p EST)

VA Loan Center Counseling
877.827.3702 (Avoid Home Foreclosures)

Vets 4 Warriors
855.838.8255 (Peer Counseling)

Veterans Families United
405.535.1925

www.militaryonesource.com
800.342.9647

Suicide Prevention Action Network
800.273.8255

Help-Yourself/Veterans
800-273-8255 – Press 1

Psychological Health Resource Ctr.
866.966.1020

Acknowledgements

Special Thanks:

Rick Lencioni, Chris White, Warren H. Chan, George Page, Jeremy McDermott, Peter Spirer, Jay Miracle, Fred Gleffe, Doug Stanton, Alex Salermon, Bob Peck, Rod Guajardo, Daniel Columbie, Angela Reyes, Jimmy Saada, Phil Saada, Rick Farina, Jay Drake, Michael Paradise, Norm Campbell, Randal Underhill, Al Dove, Jerry Austin, Michael Yon, George Jurkowski, Walter "Timmy" Flaherty, Donna Marlin, Kris "Tanto" Paronto, Walt Yost, Carl Cain, Harvey Arango, Larry Rutherford, Ron Kuvik, Jeff Burns, Korey Kiper, Joseph Carmen, Chris Mancini, George Khabbaz, Julia Mirabel, Ernie Bourcier, Kimberley Maloney, Frank Sosa, Richard "Tex" Laraway, Russel "Doc" Hall, Abe Saada, Jeanie Rinaldi, Dennis Connors, Howard Singer, Mike Thornburgh, Bobby Myers, Oscar Montoya, Brian Horowitz, Chase Bonnevile, Brandon Worthington, Arnyy Salgado, Max Furlani, Mike Furlani, Sergio Mendes, Heidi Rubio, Ryan Brady, Antoni Corone, Jason Williams, James Brusca, David Glenn, Eyal Lalo, Chris Cassidy, Fred Gleffe, Amin Ismail, Don Pegues, Tim Large, Joel Pinsky, David Scroeder, Eliad Beno, Ed Bacher, Tony Angulo, Rose Esposito, Nicole Morris, John Lally, David Cleary, Steven Kirschbaum, David Burstyn, Zelig Pinski, Adam Yuzuk, David Simoes, James McEvoy, APD Chief Steven Steinberg, Doug Jordan, Adam Sherman, Seth Kaufman, Angela Giannitti, Tom Hughes, Sam Burstyn, Anthony Carr, Tom Brunstetter, Sal Chiappetta, Ron Kastner, Ariel Weiss, Charlie Donovan, Guy Carroll, Sky Heizer, John Zimmerman, Marshall Hill, Don Sylvester, Richard Freeman, Ray Piedmont, Cindy Gutterman, Ross Marlin, Gabriel O'meara, Ken Deats, Alan King,

Michael Leoncini, Keith Goudy, Michael Yon, Patricia Cooper, ZiHong "Zee" Gorman, Jerry Pia, Leo "Rhine Wolf" Streit, City of Aventura; Miami-Dade Fire Rescue; Trinity Catholic High School; Aventura X-Storage; Pellicci's Restaurant, Stamford, CT; Subway, Aventura; Miami-Dade Fire Rescue, Lt. Felipe Lay; Aventura City Manager Eric Soroka, Tony Angulo, Nelson Reyes, Jaime Chalem, Jimmie Bernal, Janeth Carmen "Tilly" Vazquez; and APD Chief Bryan Pegues. To all of the men and women who contributed their stories, expertise, encouragement, and helped us to keep it real. Any errors are ours alone.

It should be noted that the memories of Richard's teammates in Recon's Echo Unit — Jerry Austin, Al Dove, and Stan Parker — were recorded in the book The Odyssey of Echo Company by *N.Y. Times* best-selling writer Doug Stanton. Doug was also very helpful to my documentary by providing me with photos of Flaherty that he located while researching his book. To learn more about the history of the Vietnam War go to Facebook page: Vietnam War History Org.

Lastly anyone who that knew or has further information on Richard Flaherty or pictures of him please contact us at on our website www.smallestsoldier.com or send us a message on our Facebook page The Giant Killer. We will also respect your privacy if you would like to remain anonymous email us at thegiantkiller2018@gmail.com.

David A. Yuzuk

Born and raised in Brooklyn N.Y., David was a student athlete playing football in high school and at the University of Stony Brook. David moved to Miami and was a 19-year veteran of the Aventura Police Department working as a uniformed road patrolman, undercover officer, and detective. David was awarded officer of the month on two separate occasions by his department and was recognized as officer of the month by the Dade County Chief's Association.

In 2017, he wrote and produced the documentary, "The Giant Killer" based on the life of Richard J. Flaherty. The documentary was awarded The People's Choice Award in the Silicon Beach Film Festival, Best Film in the UK monthly Film Festival, and was an official selection in the Rome International Film Festival and the Fort Lauderdale International Film Festival. David still lives in Miami and travels around the country working on various creative projects.

Neil L. Yuzuk

In 2008, After 22 years, Neil retired as a New York City Substance Abuse Prevention and Intervention Counselor for the SPARK Program at Canarsie and Franklin K. Lane High Schools.

He was happily retired working on his "Beachside PD" book series when his son, David asked for help with the documentary. "I became involved in the project early on helping to write, edit, research, photograph, make travel arrangements ... whatever else needed to be done. Most importantly, I learned about the plight of the homeless veterans and their high rate of suicide. Hopefully *The Giant Killer* will shine a light on these issues.

I am very proud to have worked on this project as we tried to uncover the riddle, wrapped in a mystery, inside an enigma—the incredible life of Richard J. Flaherty." Neil currently resides in Los Angeles where he continues to work on his writing and producing projects.